WHO'S AFRAID OF AI?

Also by Thomas Ramge

The Global Economy as You've Never Seen It
with Jan Schwochow

Reinventing Capitalism in the Age of Big Data
with Viktor Mayer-Schönberger

In Data We Trust
with Björn Bloching and Lars Luck

WHO'S AFRAID OF AI?

Fear and Promise in the Age of Thinking Machines

THOMAS RAMGE

NEW YORK

To my mother

WHO'S AFRAID OF AI?: *Fear and Promise in the Age of Thinking Machines*
Copyright © 2018 by Thomas Ramge
Translation copyright © 2019 by Thomas Ramge

Originally published in Germany as *Mensch und Maschine* by Philipp Reclam jun. Verlag GmbH & Co., KG in 2018. First published in North America by The Experiment, LLC, in 2019.

All rights reserved. Except for brief passages quoted in newspaper, magazine, radio, television, or online reviews, no portion of this book may be reproduced, distributed, or transmitted in any form or by any means, electronic or mechanical, including photocopying, recording, or information storage or retrieval system, without the prior written permission of the publisher.

The Experiment, LLC, 220 East 23rd Street, Suite 600, New York, NY 10010-4658
theexperimentpublishing.com

Many of the designations used by manufacturers and sellers to distinguish their products are claimed as trademarks. Where those designations appear in this book and The Experiment was aware of a trademark claim, the designations have been capitalized.

The Experiment's books are available at special discounts when purchased in bulk for premiums and sales promotions as well as for fund-raising or educational use. For details, contact us at info@theexperimentpublishing.com.

Library of Congress Cataloging-in-Publication Data

Names: Ramge, Thomas, 1971- author.
Title: Who's afraid of AI? : fear and promise in the age of thinking machines / Thomas Ramge.
Other titles: Mensch und Maschine. English
Description: New York, NY : The Experiment, LLC, 2019. | "Originally published in Germany as Mensch und Maschine by Philipp Reclam jun. Verlag GmbH & Co., KG in 2018"--Title page verso. | Includes bibliographical references.
Identifiers: LCCN 2018055418 (print) | LCCN 2018056654 (ebook) | ISBN 9781615195510 (ebook) | ISBN 9781615195503 (pbk.)
Subjects: LCSH: Artificial intelligence--Popular works. | Machine learning--Popular works.
Classification: LCC Q315 (ebook) | LCC Q315 .R345 2019 (print) | DDC 006.3--dc23
LC record available at https://lccn.loc.gov/2018055418

ISBN 978-1-61519-550-3
Ebook ISBN 978-1-61519-551-0

Cover and text design by Beth Bugler | Spot drawings by Dinara Galieva
Translated by Jonathan Green

Manufactured in the United States of America
First printing April 2019
10 9 8 7 6 5 4 3 2 1

CONTENTS

INTRODUCTION 1
The Kitty Hawk Moment: Why Everything Is
About to Start Happening Fast

CHAPTER 1 9
The Next Step of Automation: Machines Making Decisions

CHAPTER 2 31
Turing's Heirs: A (Very) Short History of Artificial Intelligence

CHAPTER 3 49
How Machines Learn to Learn: Artificial Neural Networks,
Machine Learning, and Feedback Effects

CHAPTER 4 63
Human Asks, Machine Answers: AI as a Daily Assistant,
Salesperson, Lawyer, and Doctor

CHAPTER 5 81
Robots as Coworkers: Smart Machines, Cobots,
and the Internet of Intelligent Things

CHAPTER 6 101
Superintelligence and the Singularity: Will Robots Seize Control?

Selected Sources 118
Acknowledgments 121
About the Author 122

> "I confess that in 1901, I said to my brother Orville that man would not fly for fifty years."
>
> —Wilbur Wright

INTRODUCTION

The Kitty Hawk Moment: Why Everything Is About to Start Happening Fast

A million-dollar prize. A one-hundred-fifty-mile route through a restricted military area in the Mojave Desert. Those were the conditions for the US Department of Defense's first Defense Advanced Research Projects Agency (DARPA) Grand Challenge for autonomous vehicles in 2004. About a hundred teams entered. The best entry ground to a halt after seven and a half miles. Eight years later, in 2012, Google issued an inconspicuous press release: Its robot vehicles, famous from YouTube videos, had covered hundreds of thousands of accident-free miles in street traffic. Today, Tesla drivers have logged more than a billion miles on autopilot. To be sure, drivers have to take control of the steering wheel every now and then in tricky situations, the arising of which the autopilot duly makes them aware. But a seemingly unsolvable problem has in principle been solved. Despite several setbacks in the form of self-driving car accidents in 2018, the path to a fully automated vehicle for the masses is merely a matter of scale and fine-tuning.

Artificial intelligence is having its Kitty Hawk moment. For decades, the pioneers of aviation promised grandiose feats, only to fall short again and again. But then the Wright brothers had a breakthrough—their first flight in Kitty Hawk, North Carolina, in 1903—and the technology took off. Suddenly, what had for years been nothing but a boastful claim now worked.

And so it is for AI: After many years of relatively slow, underwhelming progress, the technology is finally starting to perform, and now a cascade of breakthroughs are flooding the market, with many more in the works. Computer programs' ability to recognize human faces has recently surpassed our own. Google Assistant can mimic a human voice and set a haircut appointment with such perfection that the person on the other end of the line has no idea they are talking to a data-rich IT system. In identifying certain cancer cells, computers today are already more accurate than the best doctors in the world—to say nothing of average doctors working in mediocre hospitals. Computers can now beat us at the near-infinitely complex board game Go, and if that weren't enough, they've also become better bluffers than the best poker players in the world. At the Japanese insurance company Fukoku Mutual, AI based on IBM's Watson system calculates reimbursements for medical bills according to each insurance contract's individual terms. At Bridgewater, the world's largest hedge fund, algorithms do much more than merely make decisions about investments. A system fed extensive employee data has become the robo-boss: It knows what is likely the best business strategy and the best team composition for particular tasks, and it makes recommendations for promotions and layoffs.

AI is the next step in automation. Heavy equipment has been doing our dirty work for a long time. Manufacturing robots have been getting more adept since the 1960s. Until now, however, IT systems have assisted with only the most routine knowledge work. But with artificial intelligence, machines are now making complex decisions that only human beings had been able to make. Or to state it more precisely: If the underlying data and the decision-making framework are sound, AI systems will make better decisions more quickly and less expensively than truck drivers, administrative staff members, sales clerks, doctors, investment bankers, and human resource managers, among others.

By twenty years after the first powered flight at Kitty Hawk, a new industry had arisen. Soon after that, air travel fundamentally changed the world. Artificial intelligence might follow a similar course. As soon as computer programs that learn from data prove themselves more efficient at a job than people are, their dominance of that industry will be inevitable. When built into physical machines like cars, robots, and drones, they take older automation processes in the material world to the next level. Networked together, they become an internet of intelligent things capable of cooperating with each other.

Gill Pratt, head of the Toyota Research Institute, makes a historical leap even farther than the dunes of Kitty Hawk in the Outer Banks. Pratt compares the most recent advances in AI to evolutionary biology's Cambrian explosion five hundred and forty million years ago. Almost all animal phyla originated during that period, setting off a kind of evolutionary arms race as the first complex species evolved the ability to see (among other things). With eyes, new habitats could be conquered and new biological niches could be exploited. Biodiversity exploded. The emphasis on vision is important: With biological breakthroughs in digital image recognition, AI now has eyes, too, allowing it to navigate— and learn from—its environment far more perceptively. MIT's Erik Brynjolfsson and Andrew McAfee continue the evolutionary comparison: "We expect to see a variety of new products, services, processes, and organizational forms and also numerous extinctions. There will certainly be some weird failures along with unexpected successes."

AI researchers and the producers of learning software systems have a powerful current pushing them forward right now. Startups in need of capital tend to paste the artificial intelligence label on every digital application, often without any consideration of whether the system actually learns from data and examples and can extrapolate from its experiences, or whether it is de facto traditionally programmed and mindlessly follows instructions. AI sells, and many buyers—whether research sponsors, investors, or users—are able to assess a product's technical operating principles only with difficulty. A magical aura currently surrounds AI—and not for the first time.

Artificial intelligence has already been through several hype cycles. Big promises have always been followed by phases of major disappointment. During these so-called AI winters, doubts have emerged even among AI's fervent disciples over whether they were chasing pipe dreams, driven by visions inspired by the science fiction authors whose books they had devoured as teenagers.

With all this in mind, we can still safely say that research on artificial intelligence has made breakthroughs on problems it had been dashing itself against for decades. And we'd probably give AI even more credit if we didn't so often take it for granted. When a machine multiplies better than a mathematical genius, plays chess more cunningly than the reigning world champion, or reliably shows us our way through a city, we are impressed for a short time. But as soon as calculators, chess programs, and navigation apps are inexpensive products for the masses, we perceive the technology as mundane. When AI comes into its own, we have a habit of seeing rote work where we once imagined feats of intelligence.

Today, the learning curve for machines appears to be sharply steeper than it is for human beings, which is fundamentally changing the relationship between humans and machines. Euphoric utopians in Silicon Valley like the author and Google researcher Ray Kurzweil see in this the key to solving all the major problems of our time, when a wish-granting artificial general intelligence (AGI) will make our lives easier, and maybe even eternal—in the form of an upload to the cloud, as some pundits believe. Apocalypticists, who—like the Oxford philosopher Nick Bostrom—are often European, fear the seizure of power by superintelligent machines and the end of humanity. Extreme positions make good headlines. For those who advocate them, extreme positions are good business in the market for our attention. Yet these positions are nevertheless important because they are leading many people to take a closer look.

Whoever wants to explore the opportunities and risks of a new technology first needs to understand the basics. They have to find comprehensible answers to these questions: What is artificial intelligence, anyway? What is it capable of today, and what will it be capable of in the foreseeable future? And what abilities will people need to develop if machines continue to become more and more intelligent? As we find more and more precise answers to these questions, we will get ready to address the big ones: Should we be afraid of AI? Should we fear humans using AI with malicious intent? And what kind of technological framework do humans have to set in place so that thinking machines—as agents of automation—can keep their promise to make the world wealthier and safer?

> "Intelligence is what you use when you don't know what to do."
>
> —Jean Piaget, biologist and developmental psychologist

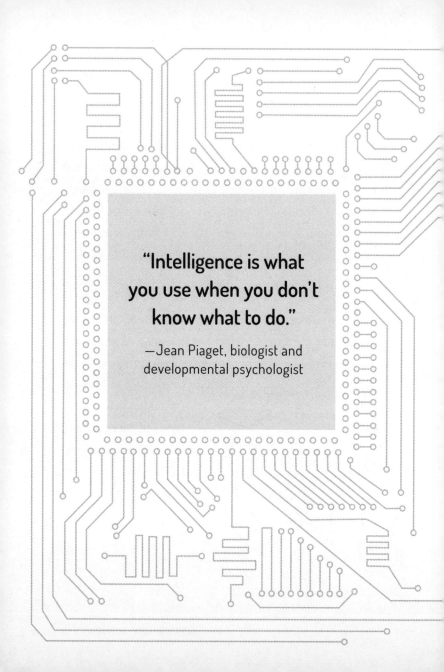

CHAPTER 1

The Next Step of Automation: Machines Making Decisions

Recognition, Insight, Action

The Tesla in autopilot is driving at eighty miles per hour in the highway's left lane. Ahead in the right lane, several trucks are driving at fifty-five miles per hour. The Tesla nears the column of trucks. The truck at the end of the convoy puts on its left blinker to signal that it wants to pass. The autopilot has to make a complex decision. Should the Tesla keep driving at the same speed or even accelerate in order to make sure it can pass the truck before the truck possibly changes lanes? Should it honk to warn the truck driver? Or should the Tesla, for safety's sake, brake and politely allow the truck to complete its passing maneuver at the cost of an increase in travel time? Braking would only be safe if there's not a lead-footed sports car driver tailgating six feet behind the Tesla, of course.

A few years ago, we would not have trusted this decision to a machine under any circumstances—and with complete justification. The technology had not yet proved that, statistically speaking, it was more likely to take us safely to our destination than we ourselves would if we were sitting behind the wheel using our own familiarity with traffic rules, knowledge based on experience, ability to anticipate human behavior, and famous gut instinct.

Today, Tesla drivers delegate many driving decisions to the computer. This is not without risks. Autonomous driving is far from perfect, whether at Tesla, Google, or the traditional car companies, which tirelessly work on autopilot systems but have not yet enabled many of their functions for safety reasons. In good weather and on clearly marked highways, today's machines are demonstrably the better drivers. It is only a question of time until this is also true in the city or at night or in fog, or until a machine decides not to drive in black-ice conditions at all because the risks are simply too high.

As the old saying in AI research goes, what's hard for people is easy for machines, and vice versa. Driving a car, which involves thousands of small but nevertheless complex decision scenarios during each trip, was previously impossible for computers. Why is that changing now? In abstract terms, it is because software that learns from data in connection with controllable hardware has increasingly mastered three core skills—recognition, insight, and implementation of an action.

In the example of the Tesla and the truck with a blinking turn signal, this means that GPS navigation, high-resolution cameras, and laser and radar sensors inform the system of even more than exactly where the car is, how fast the truck is going, the condition of the road, and if there is an emergency lane to the right. The system's image recognition software can also reliably identify that it is the truck's turn signal that is blinking and not a lamp at a construction site somewhere in the distance. Computers have gained this ability to recognize things only in the last few years. The best of them today can distinguish between crumpled paper that the vehicle can safely drive over and a rock that needs to be driven around.

All visual (and other sensory) data flow into a small supercomputer, the car's artificial brain, composed of many computing processors (called cores) and graphics processors. The processing unit has to sort the information in fractions of a second as it simultaneously synchronizes real-time data with previously collected data and rules that have been programmed into the system. The Tesla system knows that it has the right-of-way in this instance. It was equipped with the traffic rule that the truck driver is only allowed to change lanes and pass if there is no one approaching from behind. Fortified by machine learning from many billions of miles in street traffic—the so-called feedback data—the system also knows that truck drivers do not always follow traffic rules. There is a significant probability that the truck will switch lanes even though the Tesla is approaching from behind, and the car knows as well that it is not at all in the best interests of its passengers if a robot car insists upon following traffic rules when it risks a serious accident by doing so.

From the observed circumstances, programmed rules, and prior experience, the system deduces the best possibility among the many computable scenarios for avoiding an accident while still moving forward quickly. At heart, it is a cognitive decision, the choice of one course of action among many. The best solution to the problem is a probability calculation that draws on many variables.

A partially automatic assisted driving system offers its insights to the driver only as a basis for decisions, by sounding a warning beep, for example, if a truck not only signals but also makes small swerving motions indicating that the driver is truly about to turn the steering wheel to the left. The human driver can then follow the machine-derived advice or ignore it. But an autopilot worthy of the name turns its insights directly into action. It brakes or honks or drives on stoically. The computer is able to implement its decision because an autonomous vehicle is a highly developed cyber-physical system. The digital system controls the functions of the physical machine, such as the gas, brakes, and steering, with great skill. An airplane's autopilot can take off or land in normal conditions more precisely than any pilot with a captain's cap on his or her head. With a completely digital system, such as a trading bot, used for high-frequency stock market trading, implementing the decision, naturally, takes place purely digitally, but the automation principle is the same: recognize patterns in the data, deduce insights from statistics and algorithms, and implement an insight as a decision through a technological response. The machine scouts the market for trends, sees an opportunity for an advantageous trade, and clicks on "Buy it now."

Polanyi's Paradox

Measuring the effects of decisions and including the results in future decision-making is arguably the essence of artificial intelligence systems. They make decisions on the basis of feedback loops. If the Tesla in the situation described here causes an accident, it transmits this feedback back to the central computer, and all other Teslas will (hopefully) drive more defensively in comparable situations. If AI software for approving loans registers too many defaults, it will tighten the criteria for subsequent loan applicants. If a harvester receives feedback that it is picking too many unripe apples, on the next pass it will be able to make better decisions as to what ratio of red to green on an apple's surface is sufficient. The essential difference between artificial intelligence and classic IT systems lies in this ability of AI to independently improve its own calculations by classifying the results of its actions. Autocorrection is built into the system.

Since the first mainframe computers of the 1940s, programming a computer has meant that a human being painstakingly teaches a theoretical model to a machine. The model contains particular rules that the machine can apply. If the machine is fed data that fits particular tasks or questions, then it can usually solve them more quickly, precisely, and reliably than a human being can. In essence, classical programming involves transferring existing knowledge from programmers' heads into a machine. This technological approach has a natural limit: A large part of our knowledge is implicit.

It is true that we can recognize faces, but we don't know exactly how we do it. Evolution has given us this ability, but we don't have a good theory for why we are immediately able to identify Beyoncé or George Clooney, even if the light is bad and the face is half covered. It's also almost impossible to exactly describe the best way to teach a child to ski or swim. Another famous example of implicit knowledge is the answer to the question, *What is hard-core pornography?* Supreme Court Justice Potter Stewart, struggling for a legally watertight definition, found only the despairing answer "I know it when I see it." This problem has a name: Polanyi's paradox. It describes a limit that until now had seemed insurmountable for software programmers. Without theory, broken down into rules, we can't impart our knowledge and our abilities to machines.

Artificial intelligence overcomes Polanyi's paradox by having human beings create only the framework in which a machine learns how to learn. There are countless competing methods and approaches among the various schools of AI. The majority of them, however—including the most important and successful ones—follow the basic principle of giving computers not so much theories or rules, but rather goals. Computers learn how to reach these goals in a training phase involving many examples and feedback as to whether or not they have attained the goals set by human beings.

This raises the question of whether machine learning via feedback loops should be considered a form of intelligence. Many AI researchers don't especially like the concept of "artificial intelligence," preferring instead to use the designation *machine learning*.

Strong and Weak AI

The term *artificial intelligence* has been controversial since the computer pioneers associated with Marvin Minsky coined it in 1956 at their famous Dartmouth conference (more on this later). And scientists still don't even agree on what constitutes *human* intelligence. Can such a concept be appropriate for machines at all? Discussions of artificial intelligence quickly drift into very fundamental questions. For example: *Is thought without consciousness possible?* Or, *Will machines soon be more intelligent than people, and will they develop the ability to make themselves more and more intelligent, possibly developing a self-image and consciousness and their own interests in the process?* If so, will we have to grant human rights to thinking machines? Or will humans and machines just converge and form transhumanist beings escalating humanity to the next stage of evolution?

These are questions about so-called strong AI (often called general AI)—cognitively advanced, humanlike AI—and they're important. The long-term consequences of such technology should be carefully considered while it develops, not in retrospect, as in the case of, say, nuclear weapons. The last chapter of this book will touch on these questions. But these concerns are far out on the horizon. Much more urgent is the matter of weak AI, AI that is technologically possible today and in the foreseeable future. But first, let's clarify exactly what we mean by weak AI (narrow AI).

The American philosopher of language John R. Searle suggested distinguishing between strong and weak AI about four decades ago. For the time being, strong AI is science fiction. Weak artificial intelligence, on the other hand, is at work in the here and now whenever a computer system completes a task that until recently we thought only a human being exerting his or her brain in some fashion could manage—for example, case work for insurance companies, or the writing of news or sports stories.

Embedded in physical machines, AI enhances the intelligence of not only automobiles, but also factories, farm equipment, drones, and rescue and caregiving robots. But about that word *intelligence*: We can describe AI in the language we use to describe ourselves, but it's important to keep in mind that to complete a task, smart machines don't have to imitate human approaches—or the biochemical processes in the human brain in any sense at all. They typically have the ability to search autonomously for mathematical solutions, to improve the algorithms they are given, and even to develop algorithms independently. The result is that the machine does the job better, faster, and cheaper than a human being. In turn, the greater the machine's superiority compared to human problem solvers, the quicker the systems spread. This does not occur at zero marginal cost according to the principle that a digital copy costs nothing, however, as the evangelists of the digital revolution claim. Digital technology is expensive and will remain so for a while yet; ask any chief information officer if you have doubts. Hence it can be empirically proved that the cycles for introducing and disseminating new technologies are becoming shorter.

Cultural attitudes speed up or slow down the acceptance of innovations. Robots are enemies in Europe, servants in America, colleagues in China, and friends in Japan. But what universally has an effect over the long term is return on investment. Returns are often—perhaps mostly—measured in money. When Amazon invests in small, inner-city shops without human sales clerks where cameras, sensors, and RFID chips automatically add up whatever is in your shopping cart, it has to invest X million dollars in automated shelving and cash register systems, but it saves Y million dollars in personnel costs, which can be paid off in Z months or years. But when the New York Genome Center is able to analyze patients' genetic material in ten minutes using an IBM Watson application in order to suggest a likely successful therapy for an illness or injury, compared to highly qualified doctors needing one hundred sixty hours to perform the same analysis, then the return is measured not in dollars, but in human lives saved.

"Artificial intelligence will change the world like electricity did." This sentiment, or something close to it, appears in many articles and studies concerning AI. In times of technological paradigm shifts, experts' predictions—especially ones tending toward euphoria—should be treated with caution. The future can only halfway reliably be predicted from the data of the past if nothing fundamental changes.

In this respect, digitalization itself creates an interesting paradox. More data and analysis raise people's ability to forecast the future. But the disruptive nature of digital technology creates unpredictable change. And yet we are on solid ground with the hypothesis that intelligent machines will fundamentally shake up our life, our work, our economy, and our society in the coming two decades. The analogy to the introduction of electricity is correct insofar as systems that learn from data represent an integrative technology. Like the combustion engine, the development of plastics, or the Internet, it has an effect in many areas and simultaneously creates the requisite conditions for new innovations whose appearance and effects we can't even imagine today.

Electricity made possible efficient trains, the assembly line, light for libraries, the telephone, the film industry, the microwave, computers, and the battery-driven explorations of a Mars rover over rugged extraterrestrial terrain. We can't imagine modern life without electricity. Andrew Ng, a Stanford University professor and former head of the AI teams at Google and Baidu, tackles the question of which sectors AI will affect: "It might be easier to think about what industries AI will *not* transform." This is no longer a statement about the future. It describes the present, including both positive aspects and disconcerting ones.

Rage Against the Machine?

No one can reliably predict today whether artificial intelligence systems will primarily destroy human jobs or if their second wave will create new work, as has been the case in earlier technological revolutions. The machine-wrecking Luddites of the early nineteenth century smashed the first mechanical looms in central England with sledgehammers. Rage against the machine! Destroy whatever destroys you. But their rage was of little use to them. While productivity and the gross domestic product rose rapidly, for them, working conditions deteriorated. It took decades before the return on investment in automation reached their children and grandchildren in the form of higher wages and a better social safety net. The machine wreckers became a lost generation of an economic and social upheaval that the classical economist David Ricardo summarized as the "machinery question." The historian Robert Allen refers to wage stagnation from 1790 to 1840 as the almost literary "Engels's pause."

In the long term, progress found a satisfactory answer to the machinery question. The mechanization of farming replaced farmers with combine harvesters. Industrialization then not only invented the profession of the machinist who built combine harvesters, among other things, but also made use of armies of bookkeepers and later needed a great many marketing experts to take to customers the products the factories were spewing out at lower and lower prices and higher and higher quality thanks to economies of scale.

The optimistic are hoping for similar adaptations and gains at an accelerated pace—and this time without Engels's pause. They assume that learning computer systems will provide quite substantial productivity and GDP growth in the next several years, and they emphasize the opportunities for individuals, companies, and societies to invest this productivity growth in more education and better kinds of work. In a rigorously researched study, the consulting firm Accenture calculated that the American economy could grow at 4.6

percent annually by 2035 thanks to AI, almost twice as high as in a scenario without AI. In Germany, it is supposed to double the growth rate to 2.7 percent annually by 2035. Japanese politicians see artificial intelligence as a unique opportunity to get a handle on the country's demographic problems of too few workers compared to retirees. AI and robots are supposed to finally wrest the country from its stubborn stagflation.

The prospects for an economic boost from AI appear to be the best in China, however. The country has in abundance all the important components for the development and use of artificial intelligence: capital, inexpensive computing capacity, and intelligent minds that are rushing back to China from American universities and startups, and increasingly coming from China's own universities. Above all, China is the Saudi Arabia of feedback data. Its 1.4 billion inhabitants create up to half of the world's data, mainly through mobile devices, which offer even more nuanced data for mining by AI systems. From the standpoints of privacy and governmental control, this is quite worrying. From a purely economic point of view, however, it gives China the chance to further accelerate its emergence as an economic superpower and thus to lead many millions of people out of poverty.

Opposing these happy scenarios are long lists of labor economists who calculate how large a segment of the workforce can be replaced by AI. According to their gloomy prognosis, lower costs thanks to enormous economies of scale and network effects will lead to mass unemployment worldwide. In 2013, the Oxford professors Michael Osborne and Carl Benedikt Frey calculated that about half of all jobs in the United States will be seriously threatened. Their fellow economists cast doubt on the study's methodology, but it initiated a necessary debate worldwide, as it seems naive to propose that all it will take for those on the losing side of the third great wave of automation to quickly find new good jobs is a little goodwill and some government retraining programs. Many people in the United States and Europe today have the impression that digitalization is splitting the labor market into *lovely* and *lousy* jobs—pleasant and well-paid work for the highly educated winners of digitalization, especially those who build and operate the tools of data capitalism, but everyone else has to deliver packages in the rain.

This picture is surely exaggerated, but the state of affairs today is clear: Artificial intelligence and the accelerated use of robots will affect unemployment; what's uncertain is exactly what those effects will be. All forecasts—optimistic or pessimistic—have too many doubtful factors among the variables in their equations. We simply cannot estimate how well the next generations of AI systems will take over various tasks or how dynamically they will spread. The quandary regarding making a solid prognosis is at heart a question of speed. The faster AI extends itself into the human workplace, the less time remains for people to adjust their individual qualifications and their collective safety systems. A new generation of people who lose out to automation then becomes more likely. Even with all the uncertainty in the forecast, however, it is certain that politicians worldwide have until now found only a few intelligent answers to the challenges of the next major wave of automation. We are not well prepared for the return of the machinery question.

The Flaw in the Machine

The even more pressing question for humanity may well be: Will a superintelligence emerge that autonomously and in feedback loops calculates an increasingly better understanding of the world and of itself? An AI system that leads to humanity being "deposed from its position as apex cogitator," as Nick Bostrom, head of the Future of Humanity Institute at Oxford, puts it. The consequence would be that humans could no longer control the superintelligent system. And might this superintelligence even turn against humanity like in science fiction, so that in the end the machine exterminates human beings?

There's no need to delay the good news: Artificially intelligent systems will not enslave humanity in the foreseeable future. Nobody knows what computers will be able to do in two hundred years, but what we know for now is that computer scientists don't know any technological path that could lead to artificial superintelligence. The end of the world has been postponed once again. AI systems have inherent weaknesses that make them prone to making wrong decisions, limiting their use. But we must remain vigilant: The burden is on us to always critically question their algorithmic workings.

The startling thing about AI's weaknesses is how human they seem. For example, neural networks display tendencies toward prejudices that were not programmed in by the software developer, but rather emerge implicitly from the training data. If an AI-supported loan-granting process believes that it recognizes on the basis of the training data that an ethnic minority or men over 53.8 years or bicyclists with yellow helmets and eight-speed bicycles repay loans less reliably, it will take this insight into account in its scoring model—no matter if it is illegal or completely nonsensical. A learned prejudice is all the more dangerous because the machine does not reveal it. Decision-making IT systems don't tend to let slip racial slurs.

When there is a suspicion of racism, at least we human beings know what we need to pay attention to, and with good intentions we can correct it. The same problem was identified in an AI system named Compas that helps American judges decide the length of sentences, who can be released on bail, and who can be granted parole, based on factors such as the estimated risk of recidivism. The suspicion was that the system placed African Americans and Latinos at a disadvantage, and it became a textbook example of an AI system with built-in bias. In many future applications, it is possible that we would recognize a machine's prejudices too late or not at all. Who would think that a machine would discriminate against bicyclists with yellow helmets and make absurd decisions because of it?

Developers of systems for automated decision-making (ADM) have started to seek technical solutions for machine biases. IBM, for example, in September 2018 unveiled an open-source tool kit to check for prejudices in machine learning models. The application is called AI Fairness 360, and it won't be the last of its kind. AI companies know that acceptance of AI systems will heavily depend upon trust in their systems—not only in terms of fairness, but also whether they can give us good reasons to trust them.

A growing chorus of voices will continue to ask for a kind of built-in justification function for AI systems. If a machine recommends a particular chemotherapy for a certain patient, it would not be allowed to simply spit out its advice like an all-knowing oracle. It would be required to justify to the attending physician how it decided upon its result as the best solution to the problem. Such plausibility and explanatory functions already exist in rudimentary form, but they are running up against a fundamental problem.

The learning processes in neural networks are the result of millions and millions of connections, and each one of them influences the result (or more precisely the insight) a little bit. The decision-making process is therefore so complicated that the machine is unable to explain or show to human beings why it arrived at the decision "creditworthy" or "not creditworthy." This phenomenon is almost like a bad joke in the history of technology. Now it's no longer human beings who are subject to Polanyi's paradox. It's the machine that knows more than it can explain to people. This implies in turn that even if a human being notices an AI system making mistakes, they are scarcely capable of fixing those mistakes. The machine cannot show them where the mistake comes from, for it does not know the reason itself.

The human answer to the reversal of Polanyi's paradox can therefore be found only in returning to the Enlightenment's point of origin: We have to critically scrutinize everything that the machine tells us. In turning to reason and science, the Enlightenment laid the foundation for Charles Babbage's conceiving of the first computer in the middle of the nineteenth century and Konrad Zuse's building of the first programmable computer just over a century later. The connecting of computers into a worldwide network almost thirty years ago turned the gigantic digital machine into the most powerful tool that human beings have ever created. Now machines are learning how to learn—and we need more distance from them.

We have to understand when machine assistance is useful for us—and in which contexts it hinders our thinking. We will have to learn to live with so-called deepfakes that allow one to mimic another's voice or put one person's head on somebody else's body. This is already happening in pornographic videos, where the head of one person—often that of a famous celebrity—is affixed to a naked and sexually engaged body. We have to realize when AI algorithms might subvert our democracy by curating social media feeds that radicalize human political thinking in echo chambers. And politicians and citizens in democracies all over the world should follow very carefully how the military aims to use AI-powered weapons. Legislators, not people in uniforms, have to decide on all the ethical questions that arise when machines pull triggers. And developers of intelligent machines might think twice if they want to contribute to AI systems made to kill. (Many Googlers involved in AI research have decided not to, which is why in June 2018 the company declined to renew a contract with the Pentagon that had outraged many employees.)

But the fears should not blind us to the promise of AI. The automation of decision-making offers a great opportunity to individuals, organizations, and the communities that we refer to as societies. But the better that machines are able to make decisions, the more intensively we human beings have to think about which decisions we want to delegate to artificial intelligence. Because even in the age of automating decision-making with AI, it's still true that human beings have to be happy with their decisions, while computers don't. Machines will never feel what happiness is.

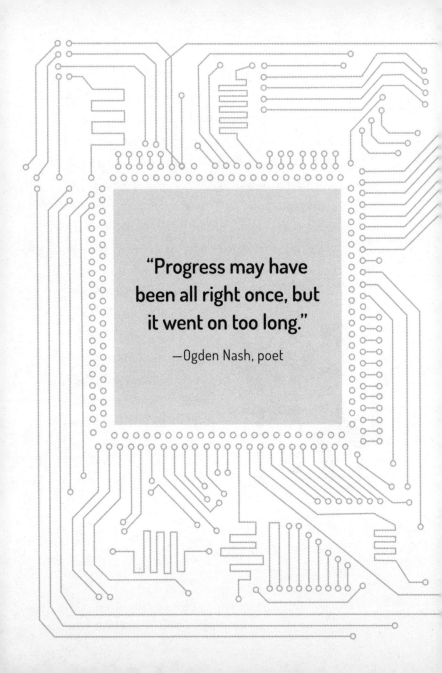

"Progress may have been all right once, but it went on too long."

—Ogden Nash, poet

CHAPTER 2

Turing's Heirs: A (Very) Short History of Artificial Intelligence

An IQ Test for Chatbots

"Can machines think?" The British mathematician, cryptographer, and computer pioneer Alan Turing posed this question in 1950 in his legendary essay "Computing Machinery and Intelligence." He gave the short answer right at the beginning of the text: Asked this way, the question can't be answered because the term "thinking" is difficult to define. Mathematician Turing had already demonstrated his practical prowess during World War II. He had decisively contributed to the first great achievement of computer science, which involved using the Bombe electromechanical machine to crack the codes of the Germans' Enigma encryption device. From that point on, the Allies were able to understand the Nazis' radio messages. Turing wanted to answer the abstract question about thinking machines with a pragmatic test: A computer should be considered intelligent if it can converse with people over an electronic connection so that the people on the other end of the line don't know if they are chatting with a human being or with a machine.

Turing had in mind that the test could be conducted by using a teletype machine as an intermediary. It was clear to this visionary of modern computer science that his experimental design was for the moment only a thought experiment—an inspiring stimulus urging researchers to teach the calculating machines of the time more than just computational routines with ever-longer columns of numbers. The first chat programs that could converse with people in a rudimentary fashion didn't appear for two more decades. But it's no accident that the question about a thinking machine appeared around the year 1950. Science and technology had made enough progress at that point in two different ways that brought machines with the ability to engage in question-and-answer games into the realm of the humanly conceivable.

In order to build an intelligent machine, you need at least two elements: a robust collection of logical rules, and a physical apparatus that can process information on the basis of these rules and draw logical conclusions from them.

From the Enlightenment to the early twentieth century, the logical foundation of classical antiquity based on Aristotle was further developed by the philosophers and mathematicians Gottfried Wilhelm Leibniz, George Boole, Gottlob Frege, Bertrand Russell, and Alfred North Whitehead. In the 1930s, Kurt Gödel demonstrated the full capabilities but also the limits of logic with his completeness and incompleteness theorems. The basic logical inventory for complex algorithms—the instructions, formulated in computer languages used by computers to carry out the tasks given to them—was thus created.

Alan Turing in turn established in 1936 that calculating machines had the fundamental capability to solve any problem that was solvable through an algorithm. His theoretical model for this was later referred to as a Turing machine. This is somewhat confusing, as the Turing machine is not a physical object, but a mathematical one. What was missing was a machine that could implement the model. It didn't have to wait long. In 1941, the German engineer Konrad Zuse made the key breakthrough. In the Z3, he created the first programmable, fully automatic digital computer, which was designed to calculate oscillations in airplane wings using a binary code of ones and zeroes. This visionary calculating machine was destroyed in an air raid in 1943.

Digital progress advanced most quickly in the United States. In 1946, the Electronic Numerical Integrator and Computer (ENIAC) was presented to the public. Researchers had been developing it since 1942 at the University of Pennsylvania, mainly to calculate artillery firing tables.

When Turing reformulated the question about thinking machines with his famous question-and-answer test in 1950, the successors to ENIAC were already reliably calculating ballistic flight paths. Provided with generous funding from the US defense budget, researchers and developers at universities and in industry rapidly improved computational performance and created the requisite hardware for the first AI program, which was introduced at the conference that gave a name to the young discipline for the first time.

Kickoff in Dartmouth

In the summer of 1956, almost two dozen mathematicians, information theorists, cyberneticists, electronics engineers, psychologists, and economists met at the Dartmouth Summer Research Project on Artificial Intelligence. The organizers had described their objective in the funding application to the Rockefeller Foundation: "Every aspect of learning or any other feature of intelligence can in principle be so precisely described that a machine can be made to simulate it."

The participants agreed that thought was possible outside of the human brain. In their view, you only had to uncover the secrets behind the brain's "neural nets," and then an electronic brain could be built. This mind-set was based on a two-hundred-year-old concept from the French philosopher Julien Offray de La Mettrie: the human being as a machine. At Dartmouth, the path to achieving it was vigorously debated for two months. Because of that continual conflict over conceptual questions, the conference attendees almost ignored Allen Newell, Herbert A. Simon, and Cliff Shaw's demonstration of a computer program named Logic Theorist that consciously imitated human problem-solving strategies—and was often able to prove mathematical theorems more elegantly than people had previously been able to. Logic Theorist was the first computer program able to process not only numbers, but also symbols and signs. It thus laid one of the most important foundations for teaching computers to understand human speech and recognize context. But even this breakthrough was hardly noticed by the researchers present. Not even the fathers of Logic Theorist recognized how visionary their program was in this respect.

Instead, trench warfare kept breaking out over the concept of artificial intelligence. The term had appeared for the first time in the conference's funding proposal, and the creator of the term, John McCarthy—a young logician and an initiator of the summer workshop along with Marvin Minsky of MIT, Claude E. Shannon of Bell Laboratories, and Nathaniel Rochester from IBM—was also dissatisfied with the concept. But especially in its abbreviated form, *AI*, the term was concise and caught on with journalists, and in the following decades it functioned superbly as a marketing label for raising research funds or investment capital for projects. Many of the participants left Dartmouth with the feeling that the substance of research had received too little attention. In retrospect, this does not change the fact that the conference was the big bang of artificial intelligence.

Back at his computer, John McCarthy developed the LISP programming language, which was soon used for many AI applications. Well-equipped institutes were founded at numerous American universities. Carnegie Mellon University in Pittsburgh, MIT, and Stanford were expanded into centers for artificial intelligence. In almost every case, participants at the Dartmouth conference headed the new institutions. The initial euphoria was followed by money. The US military and companies like IBM vigorously invested in smart computers. The research spilled into pop culture: Soon a clunky robot from Westinghouse Electric could be seen on American television telling viewers, "My brain is bigger than yours!" The first funding programs started in Europe and Japan, as well. The money was followed by the first successes, which also—or especially—impressed nontechnologists.

In 1959, Arthur Samuel wrote a program for the game of checkers that could compete with very good players. Previous checkers programs had known little more than the basic rules, and even repeatedly improved versions had stood no chance against experienced players. Samuel, an electrical engineer, made his breakthrough by teaching an IBM mainframe to play against itself and, as it did so, to record the probability that a given move in a particular situation would raise the chance of victory. With this, a human being had taught independent learning to a machine for the first time, and the approach to and concept of machine learning was born. Soon the flesh-and-blood teacher didn't stand a chance against his transistor-based pupil. This process repeated itself with many other games—chess, Go, poker—but much later than had been forecast during AI researchers' first flush of euphoria. For chess, it took until 1997. Meanwhile, the discipline was also recording successes in areas whose promise in practical value was much greater than clever computer games.

Computer Experts and Expert Computers

In 1959, the Unimate robot, which was already relatively versatile, began working on the General Motors assembly line. A little more than a decade later, Shakey, the first partially autonomous robot, which was able to explore its environment with cameras and sensors and in radio contact with a central computer, was rolling through the laboratories of the Stanford Research Institute in Menlo Park, California. In 1966, ELIZA, the first chatbot prototype with the ability to process human language, was introduced by Joseph Weizenbaum, who had been born in Berlin and fled from the Nazis to the United States with his Jewish parents. Already at this early point, ELIZA occasionally succeeded in presenting itself as a human being in short written conversations. It became famous for its doctor variant, which simulated a psychologist. Weizenbaum was surprised that many people, including his secretary, entrusted their most intimate secrets to this relatively simple program. Four years later, the expert system MYCIN was helping doctors diagnose certain blood diseases and recommending treatments. In 1971, Terry Winograd showed in his dissertation that computers were able to deduce the context of English sentences in children's books, and the first self-driving car was presented at Stanford. But these successes nevertheless fell short of the announcements and expectations.

Ever since the Dartmouth conference, AI researchers had bragged too often and too loudly. They had promised computers that would translate texts, advise customers, and take charge of administrative work on a large scale. They wanted to build smart robots that could in turn build cars to be driven by computers. They had announced that people would be able to ask a computer any question they wanted to and the computer would give the correct answer as quickly and reliably as in any science-fiction spaceship. At the beginning of the 1970s, artificial intelligence had reached the peak of overblown expectations, unable to deliver what the technologists had promised. Neither the computing nor the storage capacities were sufficient for putting the theoretical concepts into practice. The researchers weren't able to test whether their theories were suitable for practical application at all.

It became increasingly obvious that the AI community had massively underestimated the complexity of thought and language. There was a lack of all kinds of digital data that smart computers need in order to process information. Not even encyclopedias were digitized at the time. And robots first had to become much more skillful in handicraft before making them a little more clever could be contemplated. What followed was the so-called AI winter.

AI in the Hype Cycle

Government AI research programs were radically slashed. The computer industry preferred to invest in developing its hardware and in software that didn't smell of the ivory tower.

AI researchers themselves lost not only resources, but also their radiant glow as heroes of progress in information technology—and this was evidently quite beneficial for them. Many of them refocused on reaching smaller milestones. Even the terminology suddenly took a more modest turn: *rule-based expert systems* and *machine learning* didn't sound as glorious as *artificial intelligence*. But with this narrowed focus, things suddenly began to work better and better, even if they weren't announced as major revolutions that would turn the world on its head. Computers did not become superintelligent conversation partners overnight; instead, they became halfway-competent assistants for professional tasks.

Expert systems derived increasingly sensible recommendations from information such as case-based data. For this they created if-then relationships following programmed rules similar to the rule of thumb that says that if someone has a runny nose, sore throat, and fever, they have the flu, not a cold.

Building on the experience with MYCIN, expert systems came to market with more and more complex rule architectures for lung tests and internal medicine diagnosis and treatment for physicians, and the analysis of molecular structures for chemists and of rock formations for geologists. Expert systems were soon helping to configure computers and supporting employees in call centers.

The first commercial voice-recognition system came on the market in 1982. It went by the name of Covox. It didn't understand a thing, but it turned spoken language into writing relatively well. At the German Military University in Munich, the roboticist Ernst Dieter Dickmanns equipped a Mercedes van with intelligent cameras so it could drive completely independently at almost sixty miles per hour over its testing grounds.

Despite many recognizable signs of progress on a small scale, the AI winter lasted much longer than many AI researchers could have imagined it would at the beginning of the 1970s. Even in a Japan in love with robots, funding for smart machines was cut in the 1980s. And in the United States, when government and private research sponsors took stock, the result was dispiriting: little accomplished despite lofty goals. The climate for AI only warmed as the world increasingly came to be digitally networked.

With the appearance of Netscape, the first browser, in 1994, the internet was accessible to everyone, creating a space of unimagined wealth in digital data that could be processed by computers. Those computers were not overwhelmed by the newly abundant data, as the computational speed of computer chips continued to double according to Moore's law—every one to two years—while storage was becoming less and less expensive. And continually improving data connections, first by cable and then wireless, provided for better and better exchange of data.

Cloud computing finally made computing and storage capacity available worldwide like electricity from an electrical socket. Data processing on connected servers also made it possible to run complex AI applications on small consumer devices like tablets and smartphones. These technical developments were game changers for artificial intelligence.

Raw Computing Power

In the early 1990s, Polly—a likeable robot who led visitors through MIT's Computer Science and Artificial Intelligence Lab, interacted with them humorously, and simulated experiencing feelings as it did so—heralded things to come. In 1997, AI took the stage in front of a global audience: The IBM computer Deep Blue beat the reigning chess world champion Garry Kasparov. In the narrow sense, Deep Blue wasn't an artificially intelligent system that learned from its own mistakes at all, but rather an extremely fast computer that was able to evaluate two hundred million chess positions per second. The machine used so-called brute-force algorithms. The processing power was brutal, but the result seemed smart. The television images of the computer's triumph over the clever Russian captured the imaginations of AI researchers, producers, and users. Now the things that pioneers like Alan Turing had dreamed of shortly after World War II would be technologically possible: intelligent machines that could recognize pictures and people, answer complex questions, translate texts into other languages or even write creative texts themselves, pilot vehicles on land and water and in the air, predict stock prices, and make precise diagnoses when we're ill.

Jeopardy!, Go, and Texas Hold'em

In order to understand the fundamental progress of the last ten years, a glance at the competition between man and machine in the field of game AI will help. The human's defeat in chess was followed in 2011 with the victory of IBM's Watson system on *Jeopardy!* against the top quiz-show champions of recent years. Unlike Deep Blue, Watson was a system that learned from data, and its major accomplishment was not that it could look up factual knowledge from encyclopedias or newspaper articles at lightning speed; computers had been able to do that for a long time. *Jeopardy!* is a more idiosyncratic quiz show, with humorous, ironically formulated questions that require people to think outside the box. Watson's *Jeopardy!* victory therefore primarily reflected AI researchers' success with a problem that is extremely difficult for computers: semantic analysis, or the ability to understand human language and to categorize the meaning of words and sentences in the appropriate context.

Then in 2016, data scientists from Google helped an adaptive system to victory over the best Go players in the world. In this Asian board game, there are more movement possibilities than there are atoms in the observable universe. Even the fastest supercomputer is incapable of calculating them in advance, let alone a human being. So for Go, a combination of logic and intuition is required. Gifted and experienced players sense the right move in a particular situation, often unconsciously recognizing patterns that they have seen in previous games. Intuition is thus a shortcut to their experiential knowledge, which is not explicit, but rather stored implicitly in the synapses of their brains. The players cannot explain why a move seems promising. Their gut feeling leads them to make the decision.

A computer cannot feel, but it can play many millions of times against itself, just like Samuel's checkers program. Google's AlphaGo computer program built up experiential knowledge in this way in order to recognize patterns and the strategies that might fit them. To experts, AlphaGo at times seemed especially creative, with insights brought on by an ingenious mixture of pattern recognition, statistics, and random number generation. Since AlphaGo's victory, it has been clear: Intuition and creativity (depending on the definition) are no longer exclusively human domains.

Since January 2017, computers have become better bluffers than the best poker players in the world. That's when Carnegie Mellon University's supercomputer Libratus, trained by only two scientists, defeated the world's best poker players in the king of all card games, no-limit Texas Hold'em. The event generated few headlines around the world, not doing justice to its significance. For poker is a game where the qualities of an intelligent businessperson come together: strategic thinking, the ability to assess the situation and the behavior of other people, and an appetite for taking risks at the right moment. If a machine can beat people at poker, then it can also beat them at negotiations in day-to-day business.

Incidentally, for over twenty years, there has been a kind of world championship for the Turing test, the Loebner Prize. The silver medalist wins $25,000 for the system that convinces half of the jurors that it is human in a twenty-five-minute written conversation; that medal has never been awarded. Nor has the grand prize of $100,000 and a gold medal, in which systems are judged on auditory and visual communication as well as textual. Whether machines can think will probably remain a question that philosophers discuss for decades longer. But an intelligent machine that passes the test set by Alan Turing is probably only a matter of a few years away.

> "Computing machines are like the wizards in fairy tales. They give you what you wish for, but do not tell you what that wish should be."

—Norbert Wiener, mathematician

CHAPTER 3

How Machines Learn to Learn: Artificial Neural Networks, Deep Learning, and Feedback Effects

Artificial Brains?

In 1986, the American psychologists David Rumelhart and James McClelland presented a model of language learning that matched a toddler's process. To test the model, a computer was given the task of forming the past tense of English verbs. After being trained with ten commonly used verbs, eight of whose past tense was formed irregularly, the root present-tense forms—such as *start*, *walk*, and *go*—of 410 additional verbs were fed into the machine. The scientists didn't teach the system any rules, instead letting the computer discern the standard pattern for changing a present-tense verb to the past tense. The computer had to try out various options, and then received friendly feedback on its solutions and the conjugations of additional verbs as examples. But it did not explicitly receive the correct solution.

The system quickly discovered that the past tense of many verbs is formed with *-ed*, for example *started* or *walked*. Like a toddler does, it for a time overapplied that rule to verbs that did not follow it, saying *goed*, *buyed*, and *readed*. Gradually, after being instructed by its two fathers that for irregular verbs, it had to think along other lines, it began to master the conjugation of such irregular verbs—*went*, *bought*, *read*. As the program learned these exceptions, the rudimentary artificial language brain came up with the right solutions for others more quickly. After two hundred rounds, it had mastered all 420 verbs with which Rumelhart and McClelland had trained it. The computer itself had identified the rules and patterns required to complete the task assigned to it. This was precisely the point of the experiment. Naturally, it would have been easy to provide the correct answers at the start—answers for 420 verbs can be quickly programmed manually using a table. But the computer was supposed to learn how to learn through comparison and feedback, which is precisely the goal of machine learning.

People become intelligent by learning. So perhaps machines can do the same if they do indeed learn in the same way that human beings do. AI researchers made this conjecture during the first euphoric phase of the 1960s, but disillusionment soon followed. The more brain researchers discovered about the functioning of the human brain, the clearer it became to computer scientists: They could scarcely imagine how to construct an artificial brain that copied the learning processes at work inside the heads of human beings.

The human brain is the most complex structure that evolution has produced. It consists of around 86 billion nerve cells, known as neurons. On average, each neuron has more than a thousand connections to other neurons. These connections are called synapses. Neurons and synapses form an unimaginably complicated network that stores information and makes it retrievable using electrical impulses and biochemical messengers. A neuron transmits information—or more precisely an electrical impulse—to the next cell only when it has reached a certain threshold value for other cells "talking" to it within a certain period of time. Otherwise the neuron breaks the connection. This is quite similar to binary digital information processing in computers, which follows a basic principle: zero or one?

In simplified terms, information processing in our biological neural network works like this: A child sees a horse. His mother says the word *horse*. In the neural network brain, a connection is then established between the speech center and the visual center. If the child has associated the image of a horse with the term *horse* often enough, the connection becomes fixed in the brain and is always activated whenever someone says *horse* or a horse trots around the corner. Later on, probably in the second year of school, the child will create new pathways to store the fact that horse is written *h-o-r-s-e*. In a third-grade foreign-language class, neurons and synapses might link the image and the English noun with the German word *Pferd* or Spanish *caballo*. The human brain learns in a literal sense through associations and connections. The more often a connection is activated, the more it consolidates the knowledge it learns, and it corrects what it thought it knew when it receives input that this information is incorrectly wired in the brain. By linking many different connections, it can also increasingly form abstractions. Small children can recognize a comic-book mouse as a mouse, even if it wears a sombrero and has a pistol on its belt, without an older sibling having to explain it to them.

Evolution has found space for all the nerve pathways in an adult's brain—laid end to end, their length totals many tens of thousands of miles—in an average volume of less than one and a half liters. Even today, we are not even close to predicting how an artificial brain with similarly versatile properties and comparatively low energy consumption could be built. All attempts in this direction so far have failed miserably. But what machines can do very well today is to use mathematics and statistics to imitate the brain's associative learning process: the linking of spoken language, images, writing, and additional information.

The Power of the Graphics Card

The most important aid for human teachers of machines at the moment is so-called artificial neural networks (ANN). Rumelhart and McClelland used a network of this type for their verb program. But after that work was done, the approach stagnated for a long time for several reasons, including the scarcity of computers able to carry out many calculations over many nodes quickly enough. In recent years, this approach has developed rapidly thanks, among other things, to new parallel processors, the graphics processing units (GPUs) that were developed for graphics cards for 3-D computer games and then adapted to machine learning. The buzzword in Silicon Valley for this is *deep learning*. In terms of technology, deep learning stands behind most of the new applications that are labeled artificial intelligence at the moment.

ANN and deep-learning processes do not replicate the nerve pathways and electronic conducting paths of the human brain, as is often incorrectly assumed. They are, rather, statistical processes in which computer systems simulate nerve cells by using so-called nodes that are arranged in many layers behind or above each other. Typically a node is connected to a subset of the nodes of the underlying layer. The layering creates a "deep" hierarchical network. If a node is activated to a sufficient degree, it transmits a signal to the connected nodes. But just like neurons in the brain, it breaks the connection if the sum of the signals that it receives in a particular time frame falls short of a given threshold. The basic principle is the same as in the brain: If many signals arrive, they are passed onward, but when there are just a few signals, they are blocked. And like human beings, an artificial neural network learns through feedback.

The learning process—again, in highly simplified form—proceeds as follows: The computer receives the instruction to recognize horses in pictures. For this, it is first fed training data, which in this example consist of many pictures of horses that are marked as such. The network extracts a feature set of a horse's physiognomy from the data: its body shape, the positions of its ears and eyes, hooves on four legs, short coat, long tail, etc. The program proceeds layer by layer. The first layer only checks the brightness of each pixel, the next one looks for horizontal or vertical lines, the third looks for circular forms, the fourth identifies eyes, and so on. The final layer assembles the elements into a complete image. In this way, the computer produces a predictive model of how an object with the designation *horse* should look.

Like a child who is learning, the computer first has to test whether it can correctly apply the feature set. If it recognizes a horse that it has never seen before, it receives a positive response and does not calibrate its nodes any further. If it thinks a dog is a horse, then some mathematical fine-tuning is undertaken. With each iteration, the system hones its ability to recognize patterns in large data sets. This is the overarching goal of machine learning. Computer systems learn from examples, and following the learning phase, they are able to generalize their insights. The more often an algorithm has found a solution to the problem placed before it, the more accurately it can complete the task in the next round.

Supervised and Unsupervised Learning

Image recognition is only one example of applied machine learning. Neural networks are driving progress in robo-banking advice and Spotify's music recommendations. They are uncovering credit card fraud and ensuring that spam filters reject unwanted advertising. In most cases, human beings still play an important role in these systems' training phases. People have to give the systems hints on many levels so they can achieve more accurate results. Experts speak of *supervised learning*, but intelligent systems are increasingly learning unsupervised. In this case, algorithms look for patterns in data without people specifying what the algorithms should look for. The algorithms then recognize similarities and are able to automatically cluster objects—for example, finding apples in pictures without first categorizing them under the designation *apple*. Unsupervised learning is especially exciting in cases where someone does not know what he or she should be looking for.

Unsupervised learning is used, for example, in the field of IT security to defend against hacking attacks, where the goal is to discover anomalies in the operation of a company's computer network—and then immediately sound the alarm. Compared to supervised learning, however, unsupervised learning is in its infancy. Its potential is still difficult to assess, but the expectations are large. Yann LeCun, head of AI research at Facebook, maintains that "if intelligence was a cake, supervised learning would be the icing on the cake," but "unsupervised learning would be the cake" itself.

In an ideal scenario, an artificially intelligent system would on its own generate some of the data from which it learns—that is, it would be capable of deep learning. AlphaGo is an especially vivid example of this: Human beings provided the program with the rules of Go as explicit knowledge. It learned basic game skills from the many historical matches and standard situations loaded into its memory by data scientists. But that was only enough to create a Go computer at the level of a decent amateur player at best. AlphaGo became a world champion by playing against itself millions of times. With every move and countermove, it created additional data points that could be weighed in the nodes of its artificial neural network.

Feedback Creates Data Monopolies

For computer learning systems, the human platitude holds true: You never know until you try. As with people, however, it becomes true for computers only if the computer system recognizes whether its attempt succeeded or failed. Therefore, feedback data play a decisive (and often overlooked) role in learning computer systems. The more frequently and precisely a learning system receives feedback as to whether it has found the right telephone number, actually calculated the best route, or correctly diagnosed a skin condition from a photograph, the better and more quickly it learns.

Feedback is the technological core of all methods of controlling machines automatically. The American mathematician Norbert Wiener established the theoretical foundation for this—cybernetics—in the 1940s. Every technological system can be controlled and redirected according to its goals through feedback data. That sounds more complicated than it is.

Some of the first cybernetic systems were the US Army's automatic rocket-defense systems used to defend British cities against German V-1 cruise missiles. Radar detected the German rockets, informed antiaircraft cannons of the bomb's position in a continuous feedback loop, and calculated its future flight path. The cannons aimed themselves according to the continuous feedback signals and then fired at (hopefully) just the right moment. At the end of the war, the British and Americans were shooting about 70 percent of the "vengeance weapons" out of the sky.

Thankfully, feedback loops have led to more than military innovations. Without them, the Apollo missions would never have landed on the Moon, no jetliner would fly across the oceans safely, no injection pump could provide gasoline to pistons with perfect timing, and no elevator door would reopen when a human leg is caught in it. But in no other field are feedback loops as valuable as they are in artificial intelligence. They are its most important raw material.

Feedback data are at work when we begin to type a term into Google and Google immediately suggests what it presumes we are looking for. In fact, Google's suggestion might be an even better search term, because many other Google users have already given the system feedback that the term is frequently searched for when they clicked on a Google suggestion as they typed in the same or a similar search term. Then, when we accept a suggestion, we create additional feedback data. If we instead type out a different term, we do the same thing, as well. Amazon optimizes its recommendation algorithms using feedback data, and Facebook does the same for the constellation of posts that a user sees in his or her timeline. These data help PayPal predict with ever-improving accuracy whether a payment might be fraudulent; and as you can imagine, feedback about fraudulent charges tends to be quite vehement.

The sum of all feedback data has a similar effect in the age of artificial intelligence that economies of scale had for mass production during industrialization and network effects have had for the digital economy of the last twenty-five years. Economies of scale reduced the cost per item for physical products ranging from Ford's Model T, to Sony's tube televisions, to Huawei's smartphones, to a degree that Frederick Winslow Taylor, the inventor of scientific management, could hardly have imagined. The network effect—extensively investigated by the Stanford economists Carl Shapiro and Hal Varian—led to monopoly positions for digital platforms such as Amazon, eBay, and Alibaba, Facebook and WeChat, and Uber and DiDi. The network effect means that with each new participant, the platform becomes more attractive to everyone who uses it. The more people who use WhatsApp, the more users install the app, because it's more possible to contact friends or acquaintances through the app or to participate in group chats. The more smartphones that run the Android operating system, the more attractive it is for developers to develop apps for Android, again raising the attractiveness of the operating system.

The feedback effect in artificial intelligence, on the other hand, leads to systems becoming more intelligent as more people provide the machine with feedback data. Feedback data are at the center of intelligent technology's learning processes. Over the next several years, digital feedback will lead to commercially viable autonomous driving systems, language translation programs, and image recognition. And feedback data will cause lawmakers considerable headaches, because without new measures to guard against monopolies, the accumulation of feedback data over the long term will lead almost inexorably to data monopolies. The most popular products and services will quickly improve because the most feedback data will be fed into them. Machine learning will to some degree be built into these products, which means that innovative newcomers will have a chance against the top dogs of the AI-driven economy only in exceptional cases. Self-improving technology shuts out competition. Human beings will have to find a legal answer to this technological problem, a topic that we will return to in the last chapter.

> "And how does it feel to be dead?"
>
> —The ELIZA chat program, responding to the claim "I am dead."

CHAPTER 4

Human Asks, Machine Answers: AI as a Daily Assistant, Salesperson, Lawyer, and Doctor

Virtual Assistants

"Alexa, tell me a tongue twister." Alexa doesn't have to think long and says, "Bluebeard's blue bluebird." The slightly leaden woman's voice in Amazon's cylindrical speaker naturally doesn't stumble over any words. Alexa—or more precisely, the data-rich system in the Amazon cloud that's behind the Echo product family—has a large inventory of stale jokes. "How do you know if you are a pirate? You just arrrrr." The interactive speaker is also happy to sing Christmas carols on human command. Since the product's introduction in 2015, Amazon Echo's gag functions have been the subject of much laughter and much mockery, depending on one's taste in humor. But for all the debate about the system's playful functions, it's often overlooked that Amazon Echo is not a toy, but rather a technological breakthrough on the path toward intelligent daily assistants.

Simply by issuing voice commands while lying on the couch, users of Amazon Echo can turn up the heat, dim the lights, and ask Alexa to find a Netflix series that's similar to *Narcos* but not so brutal. As they stand in front of their closet, they can quickly ask how the weather's going to be, and in the kitchen, up to their elbows in cake batter, they can ask for fresh eggs to be added to the shopping list. Alexa reads the news aloud or announces when their favorite team has scored a touchdown. American customers can have their bank account balances called out to them or a pizza ordered from Domino's. Amazon's line of products are of course available worldwide, along with its well-known recommendation routines, but it would fall short of the mark to see in Alexa nothing but a sales machine. Definitions and current facts can be retrieved through dialogue. For this the system brings together information from various online sources such as Wikipedia or news websites and tries to place it in the appropriate context.

The technical term for systems like Alexa is *virtual assistant*. Often, they are simply referred to as *bots*. For several years, the giants of digital technology in the United States and Asia have been battling for dominance in speech-controlled virtual assistants. They have assembled gigantic teams of data scientists and machine-learning experts, acquired AI startups—such as Samsung's acquisition of the California company Viv Labs, progenitor of Viv, a rising star among virtual assistants—and formed surprising alliances, such as the one between Microsoft and Amazon, which in the future will let their digital helpers cooperate in service of the user. These companies are making this effort not out of pure love for technological progress, but rather out of fear for their commercial existence. Today it's clear to the strategists at Apple (with Siri), Google (with Google Assistant), Microsoft (with Cortana), and Samsung (with Bixby) that in the future access to many, probably even most, digital services will take place as it does on the Starship *Enterprise*: A human being asks, and a machine answers. If the machine hasn't mastered the game of questions and answers, the human being will look for another provider.

And users are expecting more and more precise answers to more and more complex problems. "Okay, Google, I want to fly to Switzerland for three days of skiing in March. Which ski areas are certain to still have snow, what are the nearby inexpensive hotels, when are the inexpensive flights to Zurich, and will I need to rent a car to get from the airport in Zurich to the ski resort?" For a question like that, a virtual assistant doesn't need to pass the Turing test, but rather to research and aggregate facts efficiently and present them as a foundation for decisions. In addition, there is a justified hope that we won't have to keep making a whole list of not terribly complicated but still annoying daily decisions ourselves, but can instead delegate them to intelligent machines. Virtual assistants will order printer cartridges promptly and never overlook a bill's payment deadline, but they will also recognize if the bill is too high and therefore refuse payment much more frequently than people do.

Appointment coordination assistants like Amy or Andrew, created by Silicon Valley startup x.ai, foreshadow how intelligent agents might in the future assume responsibility for irritating daily tasks. The target group is people who don't have human personal assistants. Users give these AI-assisted services access to their calendars and email accounts. Making an appointment proceeds as follows: An inquiry about a meeting arrives by email. The user signals basic agreement by email and cc's the reply to x.ai's Amy or to Julie, from AI firm Julie Desk. From this point onward, the artificially intelligent assistant takes charge of the customary email ping-pong until the parties agree on the time and place, or it's clear who will call whom at what time and at which number. In addition, more extensive systems promise to take responsibility for planning the entire daily schedule, prioritizing and if need be autonomously postponing appointments, presenting relevant information to the user in meetings that do take place, and making the user aware of anything that's been neglected.

Appointment coordination, then, is functioning quite well already. It works almost perfectly when two virtual assistants coordinate with one another on behalf of their human bosses. Computers still work best with other computers. At the same time, it's true that more and more people are listening to suggestions from computers, and not only in relatively trivial matters, such as whether it's better to wait out the traffic jam on the highway or to take the much longer alternative route over county roads—a forecasting application that Google especially is able to calculate fairly exactly and relatively easily, thanks to the abundance of real-time data from smartphones using the Android operating system.

The Sales Machine

It's no coincidence that Amazon has invested hundreds of millions of dollars in Echo's development and the system has had so much success. Since its founding in 1996, Amazon has understood, as no other company has, how to deduce the needs of its customers from data. Since the introduction of its personalized recommendation system in 1998, the company has used information about its customers to derive ever more finely grained conclusions about which products it should offer to a particular user at a particular time and at a given price in order to raise the probability that they will click on the "Add to Cart" button. Amazon, the largest online retailer in the Western world, does not provide exact numbers for how well its virtual recommendation machine works. Experts assume that the system's purchase recommendations spur about a third of all their sales. Such a high rate is possible only if customers perceive the recommendations as sensible advice and not, as is often the case, as annoying advertisements pursuing us through cyberspace to offer us products that do not interest us, or even products that we have already purchased. On the one hand, digital marketing's inane obtrusiveness has left behind vast amounts of scorched earth among customers. On the other hand, online advertising's terrible image has inspired innovative companies to try adding actual intelligence to virtual purchasing advice.

Stitch Fix is among the pioneers. The California startup offers its customers fashion by subscription—*curated shopping*, in technical jargon. It regularly sends out boxes containing five articles of clothing and accessories, and customers can keep as many as they want and return the others. So the company thrives by matching the customer's taste as closely as possible. Every return, on the other hand, costs the company money. In order to raise the hit ratio, Stitch Fix employs over eighty highly paid data scientists who use extremely complex algorithms and the newest methods of machine learning to improve forecasts for the answer to the question: Will this customer keep this article of clothing? In addition to obvious data sources such as questionnaires and the subscriber's shopping history—that is, feedback data for which clothing articles the customer has kept or sent back in the past—the system also calculates its suggestions from Instagram pictures that the customer has liked. In this way, the AI occasionally recognizes patterns in pictures that can be associated with preferences that the customer him- or herself is not aware of.

US department stores like Macy's and large grocery chains like Tesco in Great Britain or Carrefour in France are, on the contrary, trying to transfer the recommendation mechanisms proven in online commerce into the world of physical stores by using shopping assistant apps. These apps help the customer find the fastest route to the shampoo shelf if shampoo is on the shopping list, or if the customer has asked about it via a voice command while in the produce section. When a customer is standing in front of the red-wine shelf, for example, many apps will indicate without being asked that a robust Roquefort is on sale today. The problem with all these virtual purchasing advisers is that they are provided by the vendor and therefore suspected of prioritizing the vendor's interests over those of the shopper. The most-intelligent programs among the artificially intelligent shopping helpers are therefore programmed to act like reputable merchants who are interested in having a long-term relationship with a customer. They will not try to mislead customers into purchasing decisions that stand a good chance of leaving the customers annoyed afterward.

At the moment, it would be preferable if there were more vendor-independent virtual shopping assistants providing sales advice. This is the approach taken by the price-comparison apps that automatically refer consumers to sales on products they searched for some time ago but didn't buy. There is still no bot that can observe a user's consumer behavior systematically across all product categories, gradually get a better sense of the user's preferences and willingness to pay from their purchasing decisions, know that the toilet paper will be used up in a week, and also understand which routine purchases it can initiate online of its own accord and in which cases carefully prepared decision templates for the human customer are what's needed—and that ideally would even be able to negotiate the price with the vendor. For those concerned about data privacy, a virtual agent like this would represent the final step toward the transparent consumer who is vulnerable to many forms of manipulation. For anyone who doesn't like to waste time on shopping, a system like that would be a major convenience. If such an AI adviser were actually an agent of the customer and neutral toward vendors, it also wouldn't fall for dumb marketing tricks as often as we human beings do.

The Robo-Lawyer

In the field of artificially intelligent legal advice, the range of offerings is growing rapidly. What is likely the most successful virtual legal assistant in the world has the irreverent but telling name of DoNotPay. This legal-advice bot was programmed by nineteen-year-old Stanford student Joshua Browder and for now is helping its American and British users appeal parking tickets the recipient feels were issued unjustly. In a dialogue, the chatbot asks about all relevant information, and after a few minutes it spits out an individually argued, locally adapted, and legally airtight appeal.

The user then only needs to print it out, sign it, and send it off. In two years, from 2015 through 2017, this robot lawyer fended off about 375,000 penalty notices. At the moment, Browder is expanding the legal-bot's areas of expertise from traffic law to many other legal fields, such as claims against airlines, applications for maternity leave, property rental issues, and help with appealing rejected applications for asylum in the United States and Canada. Since March 2018, DoNotPay has even pushed airlines to refund money for overpriced tickets, defended the rebooking rights of passengers, and leveraged airline compliance laws against price gouging. The legal-bot is also true to its name when it comes to its fee structure: The service is free because IBM lets Browder use its Watson AI platform free of charge, among other reasons.

DoNotPay is only one example of the thousands of bots and programs engaged in legal work. There are two key drivers behind the boom in so-called legal tech. First, legal expertise is expensive, so a lot of money can be earned by either automating routine legal tasks or helping users circumvent the professionals. And second, jurisprudence is especially well suited for automation with artificial intelligence because it builds on precisely formulated rules (laws and regulations) using highly formalized language, and a large library of legal matters that are documented in case summaries and opinions, commentaries, and contracts exists that machines with pattern-recognition capabilities can draw on for comparisons. At the moment, the majority of intelligent legal tech is used by professionals—that is, by lawyers and corporate legal counsel, who use it to check over contracts for legal snares, comb through mountains of documents in conducting due diligence, and calculate the probabilities for the likelihood of success when choosing which court to file a complaint with.

But the more extensive the capabilities of legal-bots become, and the simpler their user interfaces, the more laypeople will use them directly. DoNotPay's Browder open-sourced his AI-driven chatbot technology in 2017. Any law expert, even those without technical knowledge, can now build applications. The goal is for DoNotPay to provide assistance in more than a thousand legal fields, from divorce law to private bankruptcy, quickly and without complications. That's not exactly the strong suit of every human lawyer. A free legal-bot also has no interest in formulating a contract that's as complicated as possible in order to increase its billable hours. It's also true that it may take a long time until artificial intelligence is as clever as the best and most expensive attorney in a particular field. But in standard cases, AI can already beat the average human lawyer more than occasionally—and sometimes in devastating fashion.

In a humans-versus-machines contest organized by the legal-AI platform LawGeex in February 2018, an AI system trained to evaluate contracts identified legal issues in nondisclosure agreements far more precisely than twenty experienced flesh-and-blood lawyers: 94 percent accuracy versus 85 percent. Artificial intelligence did the job faster, too—twenty-six seconds, compared to the average of ninety-two minutes it took a human lawyer.

And so the mechanisms of digital scale begin to take effect. Once AI programs have been developed and start continuously learning through feedback effects, they can be made available to many people inexpensively—at least if the providers wish to do so. As expertise is democratized, it empowers consumers and raises the competency level of the average specialist. This is also a realistic scenario for the development of artificial intelligence in the field in which machine learning has most raised the hopes of observers in recent years—medicine.

What's Wrong with Me, Dr. Watson?

Can machines diagnose human illnesses better than people can? Many experiments and studies suggest this is the case, especially regarding genetic diseases, as well as in the fields of oncology and cardiology. Thanks to deep-learning processes with CT scans, for example, tumor growth in certain kinds of breast cancers can be predicted much more accurately, making much better therapy decisions possible. But this is just the first step on the path of medical progress with AI. Using pattern recognition on cell samples, algorithms have already identified characteristics distinguishing benign from malignant tumors that had previously been unknown to the medical community.

Artificial neural networks are not just making diagnoses, but also carrying out cutting-edge research.

Many hopes are riding on the mass distribution of inexpensive sensors built into standard products that will provide data on a massive scale and thus create the foundation for AI health innovations. Smartwatches could analyze a human heartbeat around the clock and sound the alarm if a deviating pattern presages a heart attack in an individual from one particular risk group. When the cause of the arrhythmia is genetic, a person's assignment to that risk group might very well have been possible only because machine learning took place in a gene analysis process in which an unimaginable amount of genetic data has been fed into an ANN.

Based on MRI images of the brains of six-month-old babies, artificial intelligence can now predict whether they will develop autism as children or teenagers—a substantial benefit, as the earlier therapy begins, the more limited are autism's effects. AI could prospectively help not only find the best available therapy for a baby, but also develop a medication with optimal effectiveness tailored to the child's individual genome.

Researchers and startups are also working urgently on big data and machine-learning approaches that will predict the outbreak and course of an epidemic like dengue so that public-health services can introduce countermeasures in time, and in the best case confine the epidemic to the immediate outbreak location.

In short, the hope is that AI agents will dig through gene databases, patient files, scientific studies, and epidemic statistics in order to take patient care, research, diagnosis, and therapy to a new level. Naturally, it would be nice if this succeeded for as many pathologies as possible, and as soon as possible. As with all announcements of medical breakthroughs, however, caution is warranted. Researchers and company founders in medical fields tend toward exaggeration, often for marketing reasons. But an even more important reason for caution is that almost no other area of endeavor is as tightly regulated as health and medicine—from the qualifications required for medical personnel and their support staff, to the approval of medications and medical devices, to the especially strict privacy regulations for patients. There are very good reasons for this. But the cost is that for innovation, the path from the research laboratory to application in hospitals and doctors' offices is long and rocky.

The most important raw material for AI innovations in the health sector, patient data, is stored in legally sealed data silos in many different formats. For this data to become legally and technically usable by artificially intelligent applications, it usually must be laboriously anonymized and then cleaned up and homogenized.

When innovations find their way into medical practices, another fundamental question arises: Do we better trust the judgment of an artificial neural network based on data than that of an experienced doctor who might have been treating us since childhood? Maybe a computer-science student would answer with an unconditional yes. That student believes in statistics. Some patients will struggle with transferring the decision-making power from a human being to a machine.

In trend researchers' scenarios of the future, diagnostic doctors are always high on the list of skilled knowledge workers whose jobs are threatened by automation through AI. Maybe our longing for human empathy will delay their replacement. But it won't for contract lawyers, or for auditors, comptrollers, financial advisers, insurance agents, administrative officials, caseworkers, sales clerks, and—in one more irony in the history of technology—the professionals who create AI systems: programmers.

As Chapter 1 already hinted, researchers who study employment statistics and trends are skating on a very thin layer of data for their automation forecasts and the resulting negative effects on employment. Considered dispassionately, AI systems have to clear some high hurdles before people will trust their judgments and decisions. In many cases it will scarcely be possible for most nonexperts to even obtain artificially intelligent advice or to adapt it in a sensible manner without expert help. When our most valuable possession—our health—is at stake, we will hardly want to do without this adaptation. But we will require doctors to know how to use the best AI systems to prescribe therapies based on evidence and not on the basis of a gut feeling, as was all too often the case in the past.

At Stitch Fix, people still make the final decision about what items go into each box. The company's thousands of (human) stylists each enclose a personal note in their own clients' packages and are available to answer questions. Even this online pioneer of algorithmic sales consultation is convinced that in the end, people sell goods better than machines because they build human relationships with the customer.

The world might not necessarily be a worse place with fewer lawyers. What is the added value for the economy and society of ever more complicated regulations that are interpreted by ever more people? "The legal industry is more than a two-hundred-billion-dollar industry, but I am excited to make the law free," says DoNotPay's Joshua Browder. Then he adds: "Some of the biggest law firms can't be happy!" As clients, we will demand that lawyers in the future offer their services less expensively and at higher quality by using AI legal tools such as ROSS, which helps the lawyers of such major US law firms as BakerHostetler.

In almost all the knowledge-based professions in which decisions are being automated, the issue of mass unemployment for knowledge workers can also be reformulated: How will sales clerks, lawyers, and doctors ensure that they can help more people and provide better professional advice with the assistance of AI? The guiding principle here is augmented decision-making instead of pure automation. IBM CEO Virginia Rometty views things this way: "Some people call this artificial intelligence, but the reality is this technology will enhance us. So instead of artificial intelligence, I think we'll augment our intelligence." If Rometty is seeing things clearly, for knowledge workers,

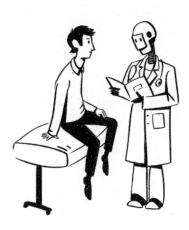

that means that artificial intelligence won't replace them in the coming years. Instead, tech-savvy salespeople, lawyers, and doctors will replace colleagues who don't know how to use AI to assist them in their decision-making.

> "I mean, being a robot's great, but we don't have emotions, and sometimes that makes me very sad."
>
> —Bender, robot on *Futurama*

CHAPTER 5

Robots as Coworkers: Smart Machines, Cobots, and the Internet of Intelligent Things

Robots on a Rescue Mission

March 11, 2011: It's a quarter to three in the afternoon when the magnitude 9.0 earthquake shakes Fukushima. At the Daiichi nuclear power plant, the external electric power supply is knocked out of service, but emergency generators kick in and the plant's three operating reactors are shut down automatically. Technical security staff report that the emergency backup batteries switch on according to plan, and sufficient coolant continues to reach the fuel rods. About an hour after the earthquake, a thirty-foot tsunami floods the inside of the reactor blocks, and a second wave follows on its heels. The seawater knocks out the backup generators, batteries, and water pumps, and the coolant in the reactor pool begins to vaporize. The fuel rods heat up to dangerous temperatures in response. Highly explosive hydrogen gas forms in the reactor containment units. The security staff desperately attempt to open vents to let the gas escape, but the vents are inoperable by remote control, and the radiation level inside the reactor building is so high that workers are unable to approach the vents to open them manually. About twenty-four hours after the initial quake, the hydrogen gas in one of the plant's six reactors explodes, followed two days later by an explosion in a second reactor and another in a third reactor on the following day. The melting of the fuel rods in these reactor blocks can no longer be stopped.

It is unclear how many people died or became seriously ill as a consequence of the worst nuclear accident since Chernobyl. More than one hundred thousand people were evacuated. The cleanup work is expected to last thirty to forty years longer and cost over $200 billion. What would have happened if the plant's engineers had been able to send robots into the overheated and contaminated reactor blocks? Would they have been able to open the vents and let the hydrogen gas escape? Would the robots have been able to take additional emergency measures to safely shut down the reactors? Could the worst-case scenario have been avoided, making the consequences of the earthquake and the tsunami much less dramatic? Not only Japanese newspaper editors asked these questions, but also the founders of the DARPA Robotics Challenge at the US Defense Department's research center.

Inspired by the recognition that robots would likely have made the situation at the Fukushima Daiichi plant much less dangerous, the goal of this follow-up competition to the Grand Challenge for autonomous driving was to give a powerful boost to the development of robotics for use in disaster control. Teams were tasked with building robots that were able to move around in a man-made environment undergoing a simulated state of emergency, with stairs to ascend, debris to climb over, and doors to open. They had to move rubbish out of their way, unplug cables, and operate tools such as drills. And of course they had to be able to open and close vents. DARPA also required the robots to be able to climb into a car and drive to the scene of the disaster.

A year after Fukushima, the first trial runs began. Finally, in June 2015, twenty-three teams from six nations met for the grand finale at a Los Angeles fairgrounds and event center. The humanoid robots on the starting line had to navigate an obstacle course that would prove them worthy—or not—of a situation similar to Fukushima. Two million dollars in prize money awaited the winning team. Thousands of spectators cheered on the metallic first responders—most weighing in at well over three hundred pounds—as if they were

Olympic decathletes participating in the king of all track-and-field events. These robots on humanitarian rescue missions did not achieve the speed, endurance, strength, and skill of those elite human athletes, however.

The results of the competition were rather mixed. Immediately after the final event, video compilations of the three competitions held over two years circulated on Twitter, Facebook, and YouTube, showing the hulking robots standing helplessly in front of a doorknob, stumbling over a few steps, or simply falling over for no discernible reason while walking along in a straight line. One robot even lost its head in the literal sense. Some observers commented that a preschooler probably could have navigated the obstacle course perfectly in less than ten minutes, except perhaps for driving the car. But it must be noted: Several robots completed all the tasks. The fastest of them, DRC-Hubo from South Korea, needed forty-four-and-a-half-minutes to finish, hardly leaving the impression that he could ever have come close to stopping the meltdowns in Fukushima. But DRC-Hubo and its fellow robots provided crucial clues for the path that technological development toward that goal might take. The frenetic cheering in the arena compared to the mockery on the Internet is an interesting reflection of the contradictory relationship that we human beings presently have with robots.

Cyber-Physical Systems

The way we imagine robots has been shaped by writers and film directors. The images in our minds are fed by director Fritz Lang's 1927 film *Metropolis* and the carefully thought-out fantasy worlds of Isaac Asimov. Robots perform their tasks with Wall-E's stylized infant face, present the most severe threats like the Terminator, or even feed romantic and erotic fantasies as the attractive Ava in *Ex Machina*. These lofty pop-culture expectations slam into the wall of technical reality when we see how slowly, clumsily, and stupidly toddler-sized plastic humanoids shove the ball toward one another at the robo-soccer world championships, and yet every year the event organizers claim that by 2050 at the latest, the reigning human world champions will stand no chance against a robot team.

At the moment, progress is being made in less spectacular fields of application than nuclear plants and soccer arenas. These venues include industrial facilities and warehouses, trains (replacing locomotive drivers), and hotel lobbies (assisting with guest check-in). Robots are taking over cleaning glass facades and carpets, picking fruits in orchards, and mowing lawns. The products of progress frequently do not look like the robots we imagine, and their developers often do not even call them robots, but rather *cyber-physical systems*. The term refers to machines in the physical world that are guided by streams of data and digital intelligence. The self-driving car is the most prominent example, but drones, intelligent milking machines and harvesters, autonomous forklifts, and smart homes are also part of the rapid process that is unifying atoms and bits.

Every year, the Amazon Picking Challenge draws major media attention. In this competition, robots have to recognize, grasp, and place into a box—undamaged—many different objects, from chocolate-chip cookies and scrub brushes to a particular book. The event is interesting to watch, but at least for the moment, machines do not stand a chance against practiced human hands. Right now, fleets of the warehouse robot Kiva have been reliably shuttling products from here to there in Amazon's logistic centers for years. Kiva has no arms and no head. It is an orange trolley, barely larger than a vacuum cleaner, with the ability to lift items. It transports bins filled with ordered goods weighing as much as three thousand pounds to packing stations, where human beings prepare them for shipment to customers. So the warehouse staff no longer have to chase through the aisles—instead, each order's items come to them, piggybacked on the flat transport robots. A central computer continuously calculates the optimal routes for the swarm based on arriving orders and directs the individual Kivas. Amazon claims that an employee can prepare two to three times as many packages per hour this way. The benefits were so convincing to Amazon that it bought Kiva Systems, the robot's developer, in 2012 for $775 million.

In the Australian and Chilean coal mines of the British-Australian mining conglomerate Rio Tinto, self-driving dump trucks from the Japanese manufacturer Komatsu and America's Caterpillar fulfill a role similar to that of the Kiva trolleys—only on a much grander scale. Trucks the size of a house, weighing over eight hundred thousand pounds, automatically drive up to an excavator, wait until they are fully loaded, and then take the raw material to a rock grinder or to loading stations for transport elsewhere.

According to Rio Tinto, the operation is about 15 percent less expensive than it would be with human drivers, and the use of robotic trucks is likely only one further step on the way to fully automated mining, similar to smart factories in which only a few people will supervise many digitally controlled machines. This process will progress especially quickly with mines, which are highly regulated areas where the same steps are repeated again and again. But even at construction sites with much more complex work processes, more and more cyber-physical systems and robots are pitching in and raising efficiency to an impressive degree.

It takes a traditional survey team about a week to survey a large construction site of forty acres. A survey drone from Aibotix, a German provider of specialty drones, can accomplish the task in a fully automatic flight lasting eight minutes—with one minor difference: The drone surveys much more precisely during its preprogrammed flight than human beings with surveyors' tools are able to.

The Australian construction robot Hadrian works much faster and also more precisely than bricklayers. In two days, it can build the shell of a single-family home with its ninety-foot gripping arm, placing brick after brick with exactly the right amount of mortar and no more than half a millimeter deviation from the 3-D blueprint that tells it where the child's room wall begins and where the kitchenette ends. Once the house is occupied, intelligent control of utilities can save between 30 and 50 percent on energy use. Smart-home systems have so far been surprisingly unfriendly for users, so the technology is spreading rather slowly. But this does not change the fact that the concept works. In intelligent houses, sensors determine if someone is in a room and adjust the lights, heat, or air conditioning as needed. A truly smart home also consults the weather service regularly and calculates how long the walls, floors, and ceilings of a well-insulated residence will continue to radiate heat. If a warm front is on its way, the system will shut off the heat earlier.

Intelligent robotization has made progress especially in agriculture, both in planting crops and in keeping livestock. *Farming 4.0* and *precision farming* are the buzzwords among tech-friendly farmers. Drones with high-resolution cameras and AI-supported automatic image analysis also play an important role, identifying where fertilizer is needed or pests have to be controlled. In Burgundy, France, the two-armed grape-growing robot Wall-Ye V.I.N. can prune up to six hundred grapevines per day, recording data about the health of the plants as it does so. Agrobots are harvesting lettuce in California and strawberries in Spain and thinning apple blossoms in Germany so the trees produce more fruit. Automatic guidance systems steer tractors and combine harvesters across the gigantic wheat fields of the American Midwest with a maximum deviation of two inches from their course, and crablike robots weighing just a few pounds are planting seedlings with an exacting eye for the seed hole in places where heavy machinery would only cause harm.

Admittedly, there have been automated feed dispensers for a long time. But the new generation of them determines the optimal amount of feed automatically by using sensor data. Milking robots not only draw milk efficiently and hygienically, but also improve the quality and quantity of milk produced by each cow by lessening the animal's stress. In addition, they are able to make inferences about the health of the animals and promptly inform farmers if a veterinarian should be called. These days, machines are getting along better and better with animals.

Human-Robot Collaboration

The Rise of the Robots, as the American futurist Martin Ford calls the current surge in automation, is in large part the result of constantly improving human-machine interaction, which owes less to humans getting better at operating the machines than to the steeper learning curve on the other side of the equation: Robots and cyber-physical systems are learning how they have to interact with us in order to be of greater benefit. Robots are turning into so-called cobots who assist us like a good coworker would.

The pioneers in putting robots to work are manufacturing plants, which have had good experiences with robots for a long time. This includes the automobile and electronics industries. Since the first wave of automation beginning in the 1960s, robots led to a strict division of labor. The at best rudimentarily intelligent machines welded and hammered away with superhuman strength behind bars and photoelectric barriers. In other zones of the facility, human beings saw to the more intricate jobs. If a human approached a machine—say, a metal sheet had slipped while being stamped and a human worker wanted to straighten it—the robot had to stop its work. Getting close to raw machine strength was much too dangerous.

For several years now, machines have increasingly been leaving their cages. They are becoming smaller, lighter, and softer than their ancestors. The LBR iiwa robot arm from the Chinese-German manufacturer Kuka, weighing only fifty-three pounds, has handed wheat beer over the counter to thirsty visitors at the Hannover industrial technology trade fair. It also washed the glasses, opened the bottles' crown caps, poured the beer, spun the bottle to dissolve the yeast, and set the perfect crown of foam on the glass with the last remnants from the bottle. The visitors did not need to be protected from the arm. If it came into contact with a person, it immediately jerked away. In addition to its high degree of dexterity, safe interaction with human beings makes a real difference.

Cobots are "social." Programmed not only to help people with specific activities, they also take note when they endanger people. For this reason, they can be integrated into work processes alongside humans, and they work hand in hand with specialists. In the BMW factory in Spartanburg, South Carolina, a cobot with the nickname of Miss Charlotte has been helping its human coworkers insert sound insulation into vehicle doors. In other factories, mobile robot arms on wheels helpfully lift heavy parts or drive in overhead screws.

For human and machine to become true partners, they have to understand each other. A number of cobots react to gestures. A wave of the hand is enough to show the bot where it should roll to. The cobots Sawyer and Baxter from the Boston firm Rethink Robotics are even learning movement sequences that a human being demonstrates for them, so a user does not need to know how to program to teach them an activity. And nonverbal communication with Sawyer and Baxter can also flow in the other direction, from machine to human being: The cobots are outfitted with a monitor at head height that displays cartoonlike eyes while in operation. Before the machine reaches in a particular direction, it points its "eyes" there, signaling its intention in the same way that human beings do, so we understand it intuitively. This humanizing of an industrial robot has the desirable side effect of improving emotional acceptance of the robot by its human coworkers—the most important prerequisite for effective human-machine cooperation. Some developers who recognize this are teaching their robots to react appropriately to human feelings.

When Robots Read Feelings

Pepper is a goggle-eyed humanoid robot standing four feet high on rollers. It has five-fingered hands and a tablet on its chest. It was developed by the French firm Aldebaran Robotics, which was absorbed in 2012 by the Japanese digital and telecommunications conglomerate SoftBank. The exceptional thing about Pepper is that the system analyzes the facial expressions, gestures, and intonation of its conversation partner and calculates how that person is feeling at the moment. If the person seems sad, Pepper sometimes performs a dance to cheer them up. Like the smart speaker Alexa or Google Home, Pepper has increasingly good conversational capabilities. The more precisely a topic is defined, the better it can answer.

Pepper offers advice for buying a smartphone at a SoftBank store, provides train schedule information for the French railway system SNCF, and acts as a tour guide on cruise ships, offering tips for life on board and short lectures about the cruise's destinations. Of course, the humanoid robot is constantly connected to the Internet in order to answer questions more proficiently by using search queries. Its manufacturer has now entered into a cooperation agreement with IBM Watson to make many of the AI platform's applications available through the robot's playful physical interface. This is intended to help with Pepper's use in schools, among other places, where it will practice math with children, quiz them on Spanish vocabulary words, and teach them calligraphy. In that environment, Pepper presents itself as patient and motivational, and it simulates empathy and rigor when the system determines they would be helpful for promoting successful learning. But the idea of empathetic robots raises some serious questions.

Simulated empathy is not empathy. In Japanese retirement and nursing homes, there are entire armies of upright-walking robots, along with nonhumanoid robots such as the fluffy seal Paro, which provides stress reduction and companionship to people suffering from dementia who may not understand that they are holding not pets on their laps, but rather machines. Does that lead to human beings taking care of these people less often in some circumstances? If futuristic robotic gloves or supportive exoskeletons enable people with movement limitations a broader range of actions, we will all welcome that development. But in delegating tasks to machines that recognize human feelings and are able to react with simulated emotions, where does the border lie? When do the advantages of improving children's learning outweigh the disadvantages of dehumanizing education? Would we rather have our personal hygiene maintained by a robot when we are old and frail, or by a human caregiver? In front of a machine, we never have to be ashamed. But can it give comfort? These are no longer theoretical questions.

The Might of the Silicon Clones

Pepper adapts its reactions to human behavior based on the cultural context. In Japan, it interacts reticently, in the United States affably, like your best buddy. That is clever and harmless. But what will happen when we can barely distinguish real people with authentic feelings from humanoid robots, like the ones Hiroshi Ishiguro, the pop star of robot developers, is attempting to create with his deceptively realistic silicon clones? Ishiguro views the humanization of robots as a necessary prerequisite for true cooperation with them. He even underwent plastic surgery so he would again resemble his own unaging duplicate like an identical twin. At the moment, both of them are traveling the world holding lectures on humanoid robotics. The audience has to guess from a distance if the real or the false Hiroshi is standing at the lectern. Is Ishiguro in the process of erasing himself through automation? Surely not. Robots are his business model. Surveyors, on the other hand, already do not stand a chance against drones.

Lawn-mowing and window-cleaning robots will make our lives easier, like the blow-dryer and the dishwasher did. Sales trends for industrial robots in the last several years have been excellent, and forecasts for the upcoming years are fantastic. The consultancy ABI Research estimates that annual sales figures will triple by 2025. The number of cobots is expected to have grown by a factor of ten between 2016 and 2020, according to Barclays. That means that the use of robots will increase to a much greater degree, because new robots typically do not replace old ones, but instead merely join the existing army of smart machines. The old ones remain in use.

The National Bureau of Economic Research calculates that each new industrial robot automates 5.6 human jobs. For Volkswagen, the biggest car manufacturer in the world, the business calculation goes like this: With the same output, one robot hour costs $3.50 to $7.00. A specialist employee costs about $60 per hour. These figures don't go unheard. In a recent study conducted by Pew Research, 72 percent of Americans are very or somewhat worried about a future where robots and computers are capable of performing many human jobs. Seventy-six percent are concerned that automation of jobs will exacerbate economic inequality, and almost as many of those interviewed, 75 percent, anticipate that the economy won't create many new well-paying jobs for those whose jobs have been automated. Sixty percent of Europeans call for a prohibition on using robots in caring for children, seniors, and the disabled. At the same time, 70 percent express a basically positive attitude toward machine assistants. According to the Pew Research study, merely a third of Americans are enthusiastic about the prospects of the age of intelligent machines. The results of these and other surveys show that our relationship to AI and robots is emotionally unsettled. Machines will have to react to this.

The transport robot Fetch has been taught by its developers to defend itself against malicious human coworkers. If rage against the machine flares up at some point and a human member of the work crew pushes or shoves the robot, its electric motors strenuously resist, making it almost impossible to push it down a flight of stairs. At the factory of the Japanese robot manufacturer Fanuc—where, incidentally, many robots are engaged in building other robots—the machines prevent such aggression through social interaction. For example, the robots are enthusiastic participants in the workplace calisthenics each morning, with human and machine both circling their arms to a musical beat. In some Japanese senior citizens' homes, the robots are not dance partners, but rather calisthenics instructors. In an aging society, there are simply not enough fitness trainers who want to take on that difficult job.

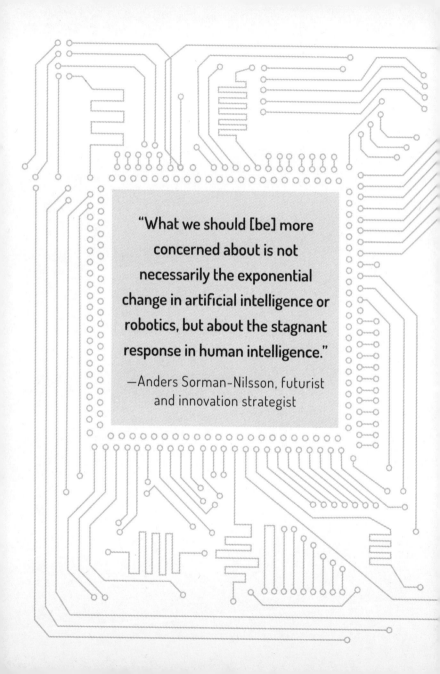

> "What we should [be] more concerned about is not necessarily the exponential change in artificial intelligence or robotics, but about the stagnant response in human intelligence."
>
> —Anders Sorman-Nilsson, futurist and innovation strategist

CHAPTER 6

Superintelligence and the Singularity: Will Robots Seize Control?

HAL Gets Serious

The crew of *Discovery 1* is irritated. On its way to Jupiter, the supercomputer HAL 9000 seems increasingly neurotic. It has clearly made a mistake in analyzing problems with the antenna module. Or is it only pretending to be mistaken? The two crew members not in cryogenic hibernation debate turning HAL off, which the artificial intelligence overhears—the astronauts do not know it can read lips. The computer decides that it must carry out its mission to Mars unswervingly and alone, and so it sees no other choice but to kill all five crew members. After doing away with four of them, it locks out astronaut Dave during a space walk. Using tricks and skill, Dave succeeds in reentering the ship through an escape hatch and advancing into the machine room. There he shuts down one computer module after another.

Hal regresses. In the end, the imbecilic artificial intelligence can only hum the children's song "Daisy Bell."

The scene is from Stanley Kubrick's film *2001: A Space Odyssey*, based on short stories by Arthur C. Clarke. The story of a malicious computer builds on an old myth in which a human being creates an artificial assistant that is supposed to serve him, but the assistant surpasses its creator by learning how to learn and finally developing its own interests and setting its own goals. Suddenly the golem is walking about. Victor Frankenstein's monster was designed as a test case to show the power of science, but he becomes the enemy. In the Terminator films, the computer system Skynet sets off a nuclear war.

Nick Bostrom, professor of philosophy at Oxford University, offers an updated version of the Frankenstein myth in his bestseller *Superintelligence*. The book is not science fiction, however; it is nonfiction. In it, Bostrom describes various scenarios for how artificially intelligent systems could become independent when they surpass the cognitive abilities of human beings. In the more harmless variation of this scenario, the process of gaining independence lasts decades or centuries. That would have the advantage of allowing people to adjust socially and culturally to the new intelligent species.

But Bostrom finds an "intelligence explosion" more probable. As soon as machines are smarter than people, they could create more and more intelligent versions of themselves within months, or perhaps even minutes. Feedback loops could lead to exponential growth in intelligence, and the philosopher speculates that the first such system would have an insurmountable head start in development. The system's *first-mover advantage* could then lead to a so-called singleton, and with it "a world order where at the highest level of decision-making there is only one agent." Bostrom finds it likely that a superintelligent system will know how to defend itself against human interference. In contrast to the inventor, Google researcher, and head of Singularity University Ray Kurzweil, Bostrom does not hold out any hope that the singleton will govern human affairs in humanity's interest and do so better than we ourselves can. The thought processes of a superintelligent machine would be, he believes, "as alien to humans as human thought processes are to cockroaches." In this scenario, the superintelligence does not even have to turn against humanity maliciously in order to threaten our existence. It would be enough if human beings were irrelevant to the omnipotent machine.

Bostrom's hypothetical horror-film predictions can at times be slightly esoteric, but his core message resonates with people who are quite familiar with intelligent machines.

Tony Prescott, a constructor of humanoid robots with a capacity for self-perception, has warned about slippery slopes in technological development. Seemingly small developmental steps could set off unforeseeable and unstoppable processes. Microsoft founder and philanthropist Bill Gates recommends reading Bostrom's book so that people can develop a feeling for the "AI control problem." Tesla founder Elon Musk thinks that AI is "far more dangerous than nukes." Together with Sam Altman, head of the startup-incubator Y Combinator, Musk founded the nonprofit organization OpenAI—buoyed by a billion dollars in funding—with the task of distributing artificial intelligence in open-source software so that it benefits rather than harms humanity.

The Intelligence Explosion and Transhumanism

Many AI researchers and developers don't buy into Bostrom's thesis; some see his most alarming predictions as, well, alarmism. And it's possible that even more AI experts regard Ray Kurzweil's thesis about a singularity being only decades away as scientifically and technologically dubious. Kurzweil assumes that computers will surpass human beings in nearly all capabilities by 2045 and that world history will then enter a phase of transhumanism. At that point, human beings will have earned the accolade of having created a godlike intelligence. Even if the research community fundamentally acknowledges that the control of AI systems is a question that scientists need to keep an eye on, most of them resist both fantasies of annihilation and techno-religious promises of salvation. They accuse both apocalyptists and euphoric utopians of understanding little about the development and difficulties of weak AI and therefore repeatedly falling for the old fantasies about strong AI. There are good reasons to remain calm—in fact many more than there are reasons to panic.

At the moment, there is no discernible development path that would make an intelligence explosion probable. Technical prerequisites for such an outcome would be exponential growth in computational performance accompanied by continuing miniaturization of computer chips. But assuming the continued applicability of Moore's law, the famous dictum according to which the computing power of integrated circuits doubles every one to two years, fails to take into account physical limitations. Conducting paths are already only a few atoms thick. Going a little smaller is probably achievable, but at some point the laws of quantum mechanics take effect and particles become confused, leaping from one conducting path to another.

Neuroscientists note that despite all the progress in AI, Pablo Picasso's bon mot about computers is still in principle true: "They are useless. They can only give you answers." Computers can apply calculation rules incredibly quickly and thus solve known problems, but they cannot identify unknown problems. They recognize patterns in gigantic amounts of data, but without data they lose all sense of direction. An important question in this context is whether computers can question rules—and thus themselves—like a critically thinking human being can. A strong artificial intelligence would have to master this trick in order to continuously reinvent itself in the same way that humans have been for thousands of years. And can machines ever really create new things? Initial forays into artificial creativity have been made, but the machine is only rolling dice to arrive at possible solutions to known problems and then asking people if the solution is any good. In this realm, it is not apparent that current AI research could endow machines themselves with the ability to be truly innovative without a human being first defining the problem. This would suggest that there might be a limit to strong AI's capacity to evolve.

The reflective and socially engaged AI grandmaster Andrew Ng gets to the heart of his skepticism about an out-of-control superintelligence with a dig at Elon Musk and his plans for a colony on Mars: "I don't work on not turning AI evil today for the same reason I don't worry about the problem of overpopulation on the planet Mars." But better safe than sorry. Google's AI unit, DeepMind, is working on concepts for built-in circuit breakers that could keep systems off the slippery slope of self-liberating information technology.

The perils of a potential strong AI are deeply concerning—it's hard to think of anything much more upsetting than the threat of the end of the human species by machine overlords—so it's no surprise that they grab most of the headlines. And, to be clear, the grave risks of superintelligence undoubtedly warrant serious consideration and sustained vigilance by AI thinkers and researchers as well as the general public. But no one knows with certainty what computers will be capable of many years from now, and we must not risk losing sight of the less dramatic but more immediate and very real dangers of the rapid development of weak AI. The most important of these hazards can be grouped under three headings: monopolization of data, manipulation of individuals, and misuse by governments. These all deserve our urgent attention, too.

Competition and Data-Monopoly Capitalism

Ever since Karl Marx, we have known that capitalism tends toward market concentration. In the industrial age, economies of scale helped large companies become ever larger.

Henry Ford showed the way. The more Model Ts he produced, the less expensively he could sell an individual car. The lower the price and the higher the quality, the more quickly his market share rose. The successful companies in the age of mass production gladly bought out their competitors in order to reap additional size advantages and at the same time to reduce competition. In the twentieth century, however, governments had an effective tool—antitrust law—to prevent monopolies (when they wanted to do so). In the age of knowledge and information—that is, since the digitalization boom of the 1990s—network effects have come more and more into play. The more customers a digital service has, the more network effects increase the service's usefulness. The operators of digital platforms in particular have succeeded in conquering market shares that the railroad barons, automobile manufacturers, and producers of instant pizzas could only dream about. In the last twenty years, corporate superstars Microsoft, Apple, Amazon, Google, and Facebook have created oligopoly structures, at times even quasi-monopolies, in the digital markets of the Western world. In Russia, Yandex dominates most digital markets. In China, Tencent, Baidu, and Alibaba have risen to become de facto monopolies with government support. US and European antitrust law is proving helpless against this. Already highly problematic today, this state of affairs will become extremely hazardous for competition as machines that learn from feedback data contribute more and more heavily to value creation. Artificial intelligence turbocharges monopoly formation because the more often products and services with built-in AI are used, the more market share they gain, and the more insurmountable their lead over their competitors becomes. In a sense, innovation is built into the product or the business process, so innovative newcomers will only stand a chance against the top dogs of the AI-driven economy in exceptional cases.

Without competition, no market economy can be successful in the long term. It eliminates itself. For this reason, Viktor Mayer-Schönberger, Oxford professor of Internet governance and regulation, and I have called for the introduction of a progressive data-sharing mandate for the goliaths of the data economy in our book *Reinventing Capitalism in the Age of Big Data*. If digital enterprises exceeded a certain market share, they would have to share some of their feedback data with their competitors—while of course adhering to privacy regulations and thus mostly in anonymized form. Data are the raw material of artificial intelligence. Only when we ensure broad access to this raw material will we make competition possible between companies and ensure the long-term diversity of AI systems. This is doubly important, because competition and diversity are prerequisites for confronting the second major danger in the age of weak AI: the manipulation or taking improper advantage of individuals through the use of artificially intelligent systems.

In Whose Interest Does the AI Agent Act?

Within a few years, we will delegate many of our daily decisions to assistants that learn from data. These systems will order toilet paper and wine exactly when we need them, because they will know how much of those things we consume. AI will organize our business trips and offer to book everything for us with a single click after we have given the itinerary a quick inspection. For lonely hearts, AI will suggest partners who are much more likely to be interesting to the lovelorn than the suggestions generated by current dating sites. But who can guarantee that the bot is really looking for the best offer? Maybe one of the oddballs on the dating site bought a premium plan and thus enjoys an algorithmic advantage. And is the self-driving taxi driving us past an electronics store because it knows that we are in the market for 3-D glasses? Maybe an ad for 3-D glasses pops up on the electronic billboard at just the moment we drive by, giving us enough time to tell the autopilot, "Stop at the electronics store for a moment!" Or would a health app raise a false alarm in order to recommend a medication?

To put it more succinctly and in somewhat more abstract terms, these scenarios raise the question: In whose interests is the virtual assistant acting? Today, most bots and digital assistants are salespeople in disguise. Alexa is built and run by Amazon, not by a startup looking on our behalf for the best deal in all online stores. This is admittedly legitimate as long as it is transparent and we are not secretly being taken advantage of. But in a world where there are many assistants, we will quickly lose track of who might be out to fool us. We will not know exactly who is advising us when we ask our smartphone or the intelligent speaker on our nightstand for advice. Often enough, we will not even care because it is so convenient, and in many cases we will even pay extra for nanny tech that pampers us.

Everyone individually will have to learn where they want to draw the line on machine-driven infantilization. We must primarily bear responsibility for our own technological self-disenfranchisement. The state and the market will need to ensure, however, that customers have access to a large selection of bots that adhere to the principle of neutrality, much as provider-independent price-comparison engines do today. There will be a need for a seal of approval—and unfairly manipulative or even fraudulent agents will have to be shut down by the government. That admittedly requires a state that is governed by the rule of law and does not itself use artificial intelligence to deceive its citizens.

The Digital Dictatorship

At the interface between the state and its citizens lurks the third and perhaps greatest danger: government misuse of weak AI for mass manipulation, surveillance, and oppression. This is no science-fiction scenario like a superintelligent computer seizing world domination and subjugating humanity. The technical possibilities for the perfect surveillance state available today read like a medley of all the political dystopian novels since George Orwell's *1984*.

The state combines surveillance cameras with automatic facial recognition and knows who crosses the street against a red light. Thanks to an autonomous drone, the surveillance camera can directly pursue the jaywalker. Voice recognition in electronic eavesdropping not only identifies who is speaking, but also determines the speaker's emotional state. AI can discern sexual preference from photos with a high success rate. Automated text analysis of social media posts and online chats can identify in real time where subversive ideas are being thought or spoken. GPS and health data from smartphones, in-app payments and credit history, digitized personnel files, and real-time criminal records provide all the information needed to calculate a citizen's trustworthiness—and of course serve up a softball for secret police to do their work efficiently. The all-powerful state naturally has social bots to disseminate personalized political messages as well.

Digital tools are not required for tyranny to exist. All the unjust regimes in world history have convincingly proved that. But in the age of intelligent machines, the question of oppression is posed with new urgency. Tech-savvy regimes are about reinventing dictatorship. In AI-powered autocracy, oppression could creep in more subtly than soldiers or police in uniform. Data show the way for the state to nudge citizens into a desired form of behavior.

China's surveillance authorities are currently building a social scoring model in which citizens' good behavior is rewarded with points. Misconduct—in private life as a result of, say, jaywalking, or in the workplace, perhaps wasting time by checking email, or politically, maybe by writing the wrong kind of post on WeChat—leads to point deductions.

Governmental guardians of public morals have access to all the data on the servers of private companies. A good score helps with promotion at work and successful application to a bank for credit. A good score favors a would-be groom in a traditional Chinese social milieu when he asks his future father-in-law for the hand of his only daughter. A low score leads to intense scrutiny by the surveillance authorities—and perhaps to interment in a prison or a work camp. The system is due to render a score for every one of China's 1.4 billion citizens by 2020.

What is astonishing from a Western perspective is that many Chinese think that the system is really not so bad, especially if they consider themselves upstanding citizens according to the government's definition and expect the AI to gain advantages for themselves. Western democracies like to take this as a particularly ominous sign of what could happen if radical parties with an authoritarian concept of statehood came to power here and gained access to AI tools for mass manipulation. And it might be even less comforting to know that the United States, Russia, Australia, Israel, and South Korea have blocked the United Nations' efforts to regulate autonomous weapons, while Russia's omnipotent president Vladimir Putin told his country's schoolchildren that "the future belongs to artificial intelligence" and then clarified: "Whoever becomes the leader in this sphere will become the ruler of the world."

A New Machine Ethics

For the time being, we don't need to be afraid of artificial intelligence running amok, but rather of malicious people who misuse it. Recent years have seen much discussion of a new machine ethics and of the question as to whether—and if so, how—ethical behavior can be programmed into machines. These debates have often been pegged to artificial dilemmas—a self-driving car is advancing toward a mother pushing a baby in a stroller, say, and a group of five senior citizens. It has to decide whom it should run over. Mother and baby, who together are expected to live another 150 years, or the five senior citizens with a collective life expectancy of 50 years? Thought experiments like this are necessary. The dignity of humanity is inviolable. In war, a general is allowed to make the trade-off of sacrificing five soldiers if he can thereby save ten. In theory, no one is permitted to do that in civilian life. In practice, a motorist driving at excessive speed with no chance to brake does make such a trade-off when choosing to drive into a group of people rather than into a concrete pillar.

The automation of decisions is of course an ethical challenge in many contexts, but at the same time it is a moral imperative. If we can cut the number of traffic deaths in half within ten years with self-driving cars, we have to do it. If we can save the lives of many cancer patients by using machine pattern recognition on cancer cells, we cannot permit progress to be slowed by a doctors' lobby that is more worried about preserving its members' co-pays. And if AI systems in

South America teach math to impoverished children, we cannot complain that it would be nicer if they had more human math teachers.

Artificial intelligence changes the fundamental relationship between human and machine less than some AI developers would like us to think. Joseph Weizenbaum, the German American inventor of the ELIZA chat program, wrote the worldwide bestseller *Computer Power and Human Reason: From Judgment to Calculation* in 1976. The book was a rallying cry against the mechanistic faith in machines that was current at that time. It deserves to be republished, at a time when belief in the technological predestination of humanity is again coming into fashion in Silicon Valley.

We can delegate decision-making to machines in many individual fields. AI systems that are skillfully programmed and fed the proper data are useful experts within narrow specialties. But they lack the ability to see the big picture. The important decisions, including the decision about how much machine assistance is appropriate, remain human ones. Or, formulated more generally: Artificial intelligence cannot relieve us of the burden to think.

The history of humanity is the sum of human decisions. We decide normatively what we want. This will remain the case. We do not even have to reinvent the positive worldview that is required for the next step in the development of the machine-aided information age: "Very simply, it's a return to humanistic values," says the New York venture capitalist, author, and TED speaker Albert Wenger. These values can in his view be expressed by a formula: "The ability to create knowledge is what makes us human beings unique. Knowledge arises through a critical process. Everyone can and should take part in this process." The digital revolution allows us to put this humanist ideal into practice for the first time in history—by employing artificial intelligence intelligently and for the good of humanity.

SELECTED SOURCES

Abbott, Ryan, and Bret Bogenschneider. "Should Robots Pay Taxes? Tax Policy in the Age of Automation." *Harvard Law & Policy Review* 12, no. 1 (2017): 145–75.

Bostrom, Nick. *Superintelligence: Paths, Dangers, Strategies.* Oxford, UK: Oxford University Press, 2014.

Brynjolfsson, Erik, and Andrew McAfee. *The Second Machine Age: Work, Progress, and Prosperity in a Time of Brilliant Technologies.* New York: W. W. Norton, 2014.

Brynjolfsson, Erik, and Andrew McAfee. "The Business of Artificial Intelligence." *Harvard Business Review*, July 2017. hbr.org/cover-story/2017/07/the-business-of-artificial-intelligence.

Ford, Martin. *Rise of the Robots: Technology and the Threat of a Jobless Future.* New York: Basic Books, 2015.

Frey, Carl Benedikt, and Michael A. Osborne. *The Future of Employment: How Susceptible Are Jobs to Computerisation?* Oxford, UK: Oxford Martin School, University of Oxford, 2013. oxfordmartin.ox.ac.uk/downloads/academic/The_Future_of_Employment.pdf.

Friend, Tad. "How Frightened Should We Be of AI?" *The New Yorker*, May 14, 2018. newyorker.com/magazine/2018/05/14/how-frightened-should-we-be-of-ai.

Husain, Amir. *The Sentient Machine: The Coming Age of Artificial Intelligence*. New York: Scribner, 2017.

Kurzweil, Ray. *How to Create a Mind: The Secret of Human Thought Revealed*. New York: Viking, 2012.

Lotter, Wolf. "Der Golem und du" [The Golem and You]. *brand eins*, July 2016. brandeins.de/archiv/2016/digitalisierung/einleitung-wolf-lotter-der-golem-und-du.

Mayer-Schönberger, Viktor, and Kenneth Cukier. *Big Data: A Revolution that Will Transform How We Live, Work, and Think*. Boston: Houghton Mifflin Harcourt, 2013.

Mayer-Schönberger, Viktor, and Thomas Ramge. *Reinventing Capitalism in the Age of Big Data*. New York: Basic Books, 2018.

Ng, Andrew. "What Artificial Intelligence Can and Can't Do Right Now." *Harvard Business Review*, November 9, 2016. hbr.org/2016/11/what-artificial-intelligence-can-and-cant-do-right-now.

Ramge, Thomas. "Management by Null und Eins" [Management by Zeros and Ones]. *brand eins*, November 2016. brandeins.de/archiv/2016/intuition/intuition-im-management-by-null-und-eins.

Standage, Tom. "The Return of the Machinery Question."
Special Report: Artificial Intelligence. *The Economist*, June 2016. economist.com/sites/default/files/ai_mailout.pdf.

Shapiro, Carl, and Hal R. Varian. *Information Rules: A Strategic Guide to the Network Economy.* Boston: Harvard Business School Press, 1999.

Solon, Olivia. "More Than 70% of US Fears Robots Taking over Our Lives, Survey Finds." *Guardian*, October 4, 2017. theguardian.com/technology/2017/oct/04/robots-artificial-intelligence-machines-us-survey.

Tegmark, Max. *Life 3.0: Being Human in the Age of Artificial Intelligence.* New York: Alfred A. Knopf, 2017.

ACKNOWLEDGMENTS

I wrote this book shortly after having finished the manuscript of *Reinventing Capitalism in the Age of Big Data*, which I had the honor of coauthoring with the Oxford professor Viktor Mayer-Schönberger. Many thoughts from the discussions I had with you, dear Viktor, have found their way into this book. Thanks once again!

The mathematician, neighbor, friend, and tennis partner Max von Renesse has inspected the more mathematical and technical sections of the text. Thank you, dear Max, for your valuable, scientifically sound advice on how to think and write more accurately.

I am very grateful to the American Germanist scholar Jonathan Green for supplying additional content-related feedback when translating the German text into English.

Nicholas Cizek, thanks again for trusting (and investing) in my writing and for your excellent editing.

And I thank Lisa Adams of the Garamond Agency for taking care of all transatlantic legal and tax stuff, the kinds of things intelligent machines won't be able to sort out for us for many years—maybe many decades.

Dear Anne, dear Moritz, Dankeschön, once more, for your love and patience.

ABOUT THE AUTHOR

THOMAS RAMGE is the author of more than a dozen nonfiction books, including *Reinventing Capitalism in the Age of Big Data*, coauthored with Viktor Mayer-Schönberger, and *The Global Economy as You've Never Seen It*, written with Jan Schwochow. Ramge has been honored with multiple journalism and literary awards, including the Axiom Business Book Award's Gold Medal, the getAbstract International Book Award, *strategy+business* magazine's Best Business Book of the Year (in Technology and Innovation), the Herbert Quandt Media Prize, and the German Business Book Prize. He lives in Berlin with his wife and son.